W9-CDH-487

The ULTIMATE CLOCK BOOK

The ULTIMATE CLOCK BOOK

40 Timely Projects from Wood, Metal, Polymer Clay, Paper, Fabric, and Found Objects

Paige Gilchrist

CONSULTING DESIGNERS: ALLISON AND TRACY PAGE STILWELL

A DIVISION OF STERLING PUBLISHING CO., INC.
NEW YORK

Editor: **Paige Gilchrist**
Art Direction and Production: **Celia Naranjo**
Photography: **Evan Bracken**
Illustrations: **Bernadette Wolf**
Editoral Assistance: **Heather Smith and Catharine Sutherland**
Production Assistance: **Hannes Charen**

Library of Congress Cataloging-in-Publication Data

Gilchrist, Paige.
The ultimate clock book: 40 timely projects from wood, metal, polymer clay, paper, fabric, and found objects/Paige Gilchrist; consulting designers, Allison and Tracy Page Stilwell.
p. cm.
ISBN 1-57990-166-2
1. Handicraft. 2. Clock and watch making. I. Stilwell, Allison. II. Stilwell, Tracy Page. III. Title.

TT157 .G46 2000
681.1'13—dc21 00-028232

10 9 8 7 6 5 4 3 2 1

Published by Lark Books, a division of
Sterling Publishing Co., Inc.
387 Park Avenue South, New York, N.Y. 10016

Distributed in Canada by Sterling Publishing,
c/o Canadian Manda Group, One Atlantic Ave., Suite 105
Toronto, Ontario, Canada M6K 3E7

Distributed in Australia by Capricorn Link (Australia) Pty Ltd.,
P.O. Box 6651, Baulkham Hills,
Business Centre, NSW 2153, Australia

If you have questions or comments about this book, please contact:
Lark Books
50 College St.
Asheville, NC 28801
(828) 253-0467

Manufactured by Dai Nippon in Hong Kong

ISBN 1-57990-166-2

Table of Contents

Clock Watching

Some economists say the typical worker's time at the office has increased dramatically over the last quarter century. Others argue that we actually spend less time on the job. We just feel busier because technology gives us access to our work 24 hours a day—and we're juggling so much else.

Most of us don't have time to figure out whose statistics to believe.

We're conducting our own research in living laboratories: pulling together the details for Saturday's slumber party during lunch hour on Friday, getting the headlight fixed on the way to the meeting, and figuring out whether we can squeeze in a stop at the supermarket between yoga, helping with homework, and getting the report in on deadline. The startling results? There aren't enough hours in the day.

It would be far too ambitious of us (or way too trite) to suggest that a book on crafting clocks is your answer to making more time in your day. There are plenty of guides on the market if that's what you're after—books that promise to teach you how to transform 24 hours into 48, how to train your dog to take less of your time (fewer stops at fire hydrants is one top tip), and (while you're at it) how to find more time for saving the planet. Whew!

We don't get into doubling hours, speed-walking dogs, or saving the world. What we do offer—in the most literal sense—are new ways to look at time. Forty new ways, in fact!

We asked a talented team of designers to create the most clever clocks they could imagine. What follows is the terrific collection of projects they came up with. Some are as simple as drilling a hole and adding a clock movement to an easy-to-find (but often unexpected) item (see the fabulous vintage pocketbook clock on page 103, for example). Others are not only quick, but easy to adapt when you want a new look (check out the clock made from a paper lunch bag on page 74). Many more involve simple painting, woodworking, and other general craft techniques and result in clocks that will stand the test of time.

A few designers went whimsical, adding wings to furniture pieces (time flies, of course) or equipping Alice's White Rabbit from Wonderland with a working pocket watch. Others incorporated coiled copper or salvaged metal into inventive modern designs. The rest worked with materials ranging from weathered barn wood and polymer clay to family photos and decorative handmade paper to create imaginative new ways of viewing time.

So let the economists duke out the details about how much time we really have. Most of us already know the answer: not enough. Maybe you can't change the fact that you've always got to keep one eye on the clock. But with this book as your guide, you *can* make sure it's a clock you like looking at.

TIMELINE

A History of Keeping Track of Time

SINCE THE BEGINNING OF TIME, we humans have been preoccupied with measuring and recording its passage. Tens of thousands of years ago, the absence of back-to-back meetings and social obligations may have helped people feel they had a bit more time on their hands.

A modern reproduction of an 18th-century candle timepiece; an oil lamp clock, ca. 1800; and a sand clock that measures one hour, ca. 1640

As a result, they weren't obsessed with managing it down to the minute; they were more interested in tracking sweeping progressions—the change of seasons, the movements of celestial bodies, and such.

Ice-age hunters in Europe, for example, scratched lines in sticks and bones, keeping count, historians speculate, of the days between phases of

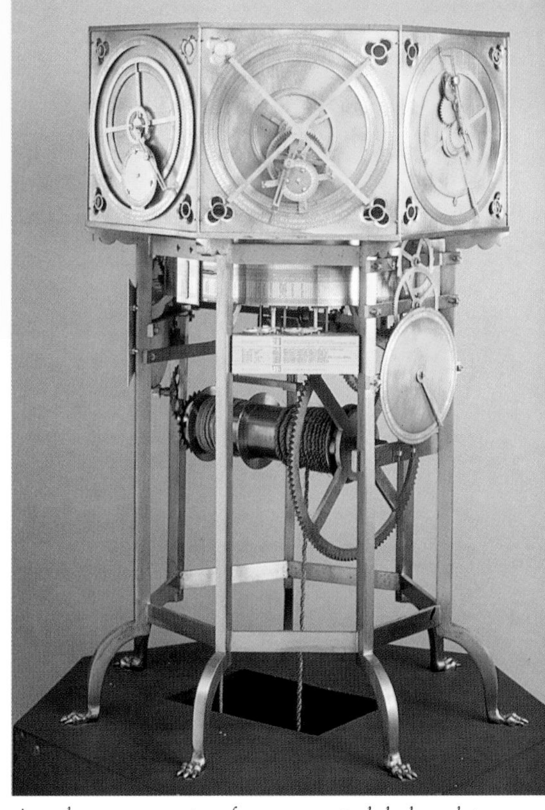

A modern reconstruction of an astronomical clock made in Italy in 1364. The elaborate system shows the position of the sun, moon, and the five planets then known.

the moon. The alignments of Stonehenge, built more than 4,000 years ago in England, indicate that part of its purpose was to determine when seasonal and celestial events, from solstices to lunar eclipses, would occur. Later, Egyptians, Sumerians, Babylonians, and Mayans all developed celestial-cycle calendars, some similar to

This Renaissance table clock was made in Germany in 1573.

those we follow today, with 29- and 30-day months and 365-day years.

As civilizations developed, people became interested not only in what time of year it was (and, therefore, when to plant seeds and harvest crops) but also in how to organize each day. Calendars were no longer enough. They needed clocks. Those who invented the earliest ones determined that a system needs only two components to qualify as a clock: a regular, repetitive action and a way of monitoring and recording that action.

At least as far back as 3500 B.C., sky gazers in Egypt were using the motion of the sun across the earth's sky as the action and measuring it by tracking the shadows it cast. Based on those measurements, they divided their days in half. The moving shadows of tall, slender obelisks served as rudimentary sundials at first. Later, shadow clocks, with markers for units somewhat like our hours, allowed Egyptians to divide days into even smaller sections. And so began the tradition of timekeeping—and the ongoing search for more consistent repetitive actions and more precise ways of measuring them.

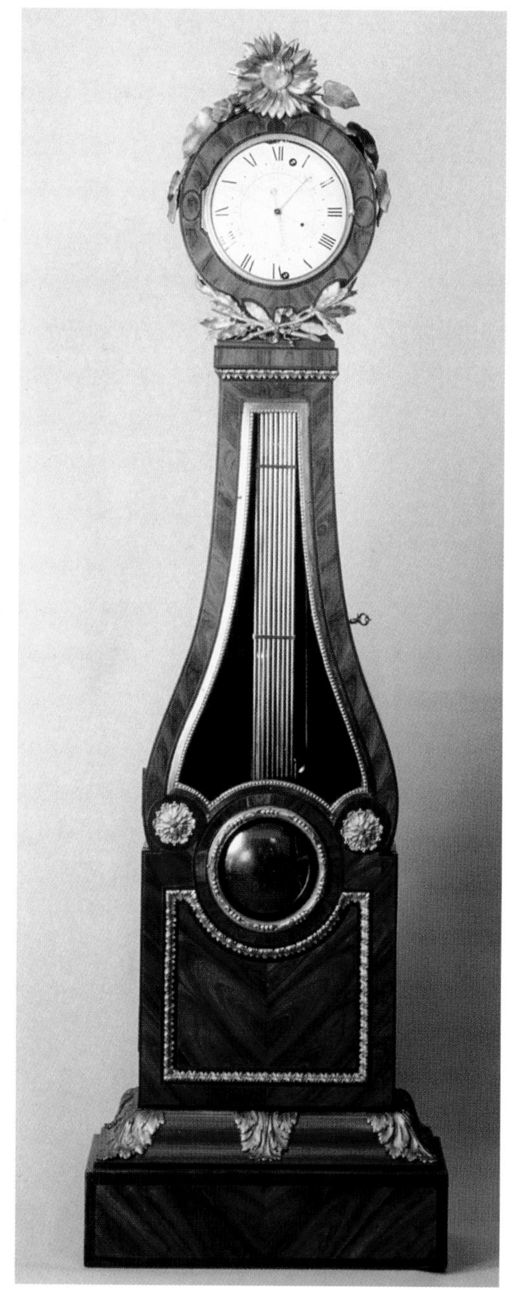

Louix XV French long-case clock, made in Paris, ca. 1770

Since the earth's motion as it orbits the sun isn't uniform, the time it took for a shadow to cross a mark on an Egyptian sundial wasn't necessarily the same from day to day or season to season. More important, sundials won't give you the time of day in the middle of the night or on a cloudy afternoon. People turned to alternatives such as candles marked in increments, oil lamps with marked reservoirs, sand-filled hourglasses, and incense that burned at a regular pace.

Water clocks were the most widely used early timekeepers that didn't rely on movements of celestial bodies. They were developed around 1500

The birds inside the fire-gilt bronze and brass cage of this French clock, ca. 1780, move their heads, wings, and tails every hour or "upon request."

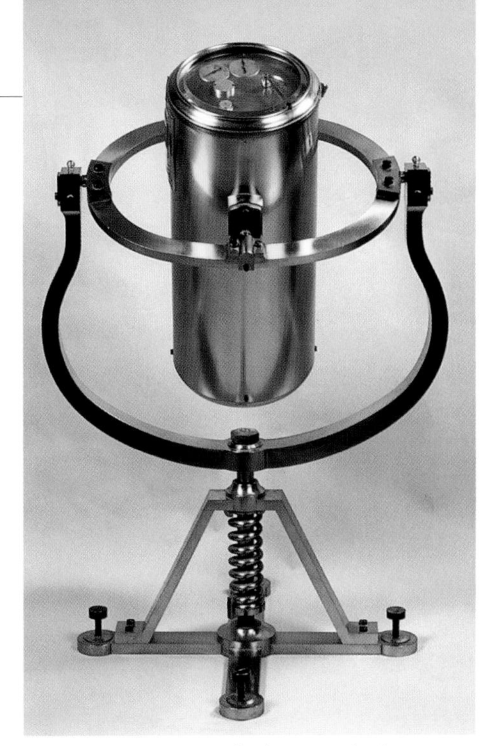

Made in Paris in 1776, this large, weight-driven marine chronometer features separate dials for hours, minutes, and seconds.

B.C. and popularized by the Greeks, who called the stone vessels that dripped water at a constant rate clypsedras or "water thieves." As Greek, Roman, and later Chinese clypsedras became more elaborate and mechanized, they rang bells and gongs, opened doors and windows to reveal tiny figures, and moved pointers, dials, and astrological models of the universe to indicate certain hours of the day.

While the pageantry of water clocks was nice, users learned that water flow is difficult to regulate. Demands for more accurate timekeeping continued. By the mid-14th century, mechanical clocks whose gears were driven by weights rather than water began appearing in towers throughout Europe. The first one publicly sounded the hour in Milan, Italy, in 1335. In the early 1500s, they became even more accurate, when springs replaced the heavy and irregular drive weights.

Later that century, a daydreaming Galileo Galilei, while gazing toward the heavens in the Cathedral of Pisa, made a discovery that would eventually take clocks to a new level. Watching the slow swing of bronze lanterns suspended from the ceiling, he happened upon the principle that led to the inven-

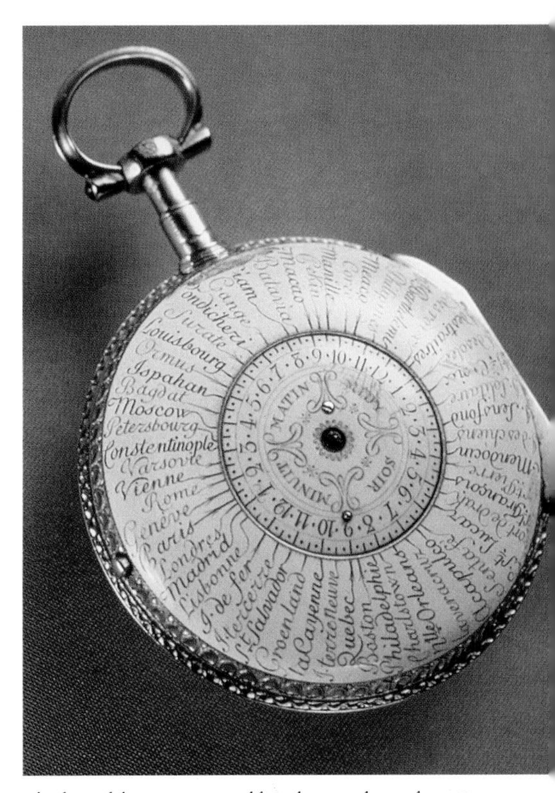

The face of this 18-carat-gold pocket watch, made in Geneva, Switzerland, ca. 1780, displays the names of 53 towns. The center section of the dial moves counterclockwise once every 24 hours, thus indicating the local time for each town.

Clock in a vase made of Niderviller china, France, ca. 1780. The dial consists of two pivoting enameled rings: the lower with Roman numerals and quarter-hour markings, and the upper with Arabic numbers for the minutes.

tion of the pendulum and vastly improved the accuracy of mechanical clocks. By timing the movements of the lanterns against his pulse, Galileo noticed that identical lanterns suspended the same distance from the ceiling always took the same amount of time to swing back and forth. Just before his death, he proposed applying the principle to the workings of clocks.

Dutch scientist Christiaan Hugens did just that in 1665, when he created the first clock with a pendulum. Its slow, steady swing made the clock more accurate than any mechanical clock before it, with an error rate of less than one minute a day. Others continued to improve on the accuracy of pendulum clocks for centuries to come by freeing the main pendulum from mechanical tasks that would disturb its regularity.

In the 1930s, the invention of quartz clocks began to solve the problem altogether. Because they operate according to crystal vibrations (which have extremely constant frequency) rather than the more irregular

This Renaissance neck watch from France, ca. 1590, features a sundial, a mechanical watch, and an alarm.

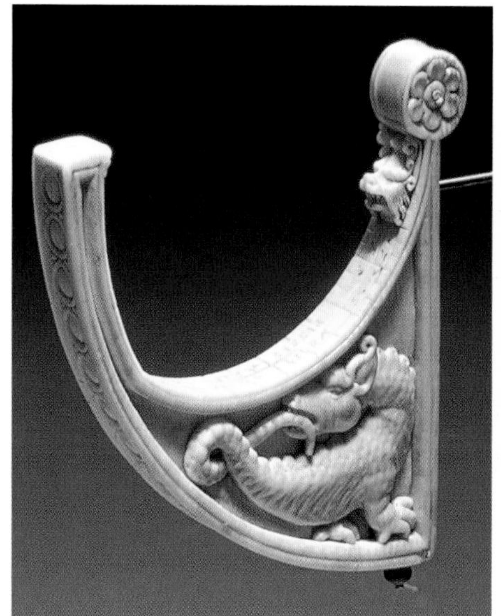

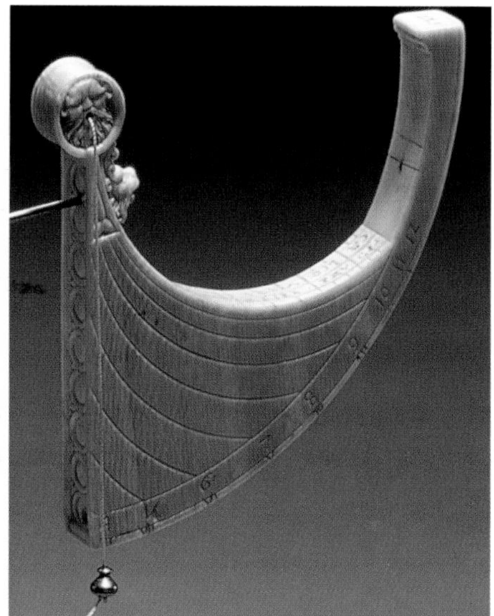

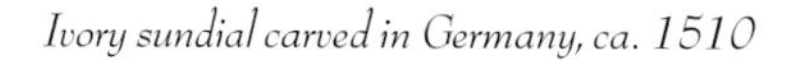

Ivory sundial carved in Germany, ca. 1510

turning of pendulum-driven gears, their timekeeping accuracy is very high (to within one ten-thousandth of a second over a period of months). Quartz clocks, now widely available and quite inexpensive, have replaced mechanical clocks today as the standard for general timekeeping. They're what most of us refer to if we need to know how much time we've got before the movie starts or whether we're late for work. Still, for scientific experiments, high-tech research, or anything else that requires the most accurate and stable timekeeping measures available, the ultimate device is an atomic clock.

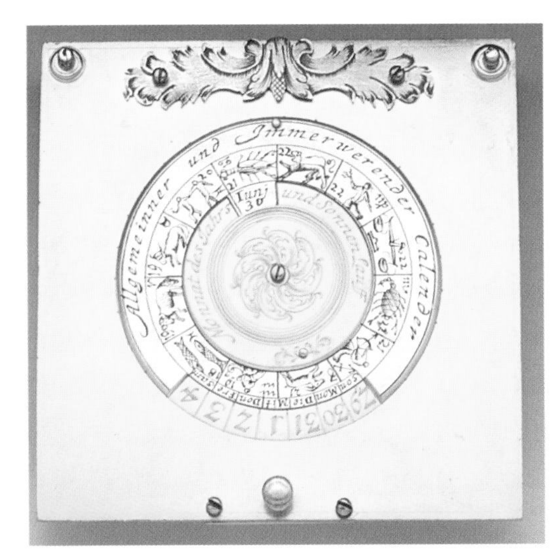

Travel sundial (front and reverse) from Germany, ca. 1700

Atomic time, adopted in 1972 as the primary reference for all scientific timing, measures the vibrations of atoms. If you need precision, atomic time is it; it defines a second as 9,192,631,770 oscillations of the atom cesium-133. Better yet, atomic clocks have long-term accuracy, losing only about one-thousandth of a second every 300 years. And they're getting more exact. Scientists say that soon atomic clocks will use lasers rather than magnets to monitor atomic vibrations, allowing them to measure time with an error rate of less than one second every 12 to 18 billion years.

The very thought is enough to leave those who tend to run a bit late—then blame it on a slow-moving clock—longing for a system of crudely carved sticks that tracks nothing more than the next full moon.

Made in France, ca. 1770, this table clock features a cut-glass globe housing a planetarium. The rotation of the planets is powered by the clock movement.

BASICS

THE MODERN CRAFT OF CLOCK MAKING welcomes anyone who can drill a hole, insert the shaft of a quartz movement, and screw a set of hands in place.

Let's be clear. As clock crafters, we're a different lot from traditional clock makers (also known as horologists), who must study for years before mastering the science of making clocks tick. They know how to piece together intricate mazes of tiny gears and springs and end up with a system that runs like...clockwork. Thanks to modern technology, that's specialized knowledge the rest of us don't have to have.

Timekeeping systems are now packaged in inexpensive, ready-to-go kits. For less than the cost of a casual lunch out, any craft store will sell you a single, small package that contains everything you need to make your clock run.

THE MOVEMENT is the mechanism that keeps precise time and moves a clock's hands around its dial in accurate, measured units. Those packaged for craft projects are battery-operated quartz movements. They come in a variety of sizes, with larger movements providing better support for longer and/or heavier hands.

THE SHAFT (sometimes also called the spindle) sticks out from the middle of the movement. You'll insert it from the back of the clock, through your drilled hole, and up into the clock face, then fix the hands in place on the part of the shaft that protrudes through the face. (All you need to do this are a few washers and nuts, which are typically packaged with the movement.) Shafts come in various lengths to fit clocks of different thicknesses, so it's best to know how thick your clock will be before purchasing your movement.

Movements with shafts of various lengths

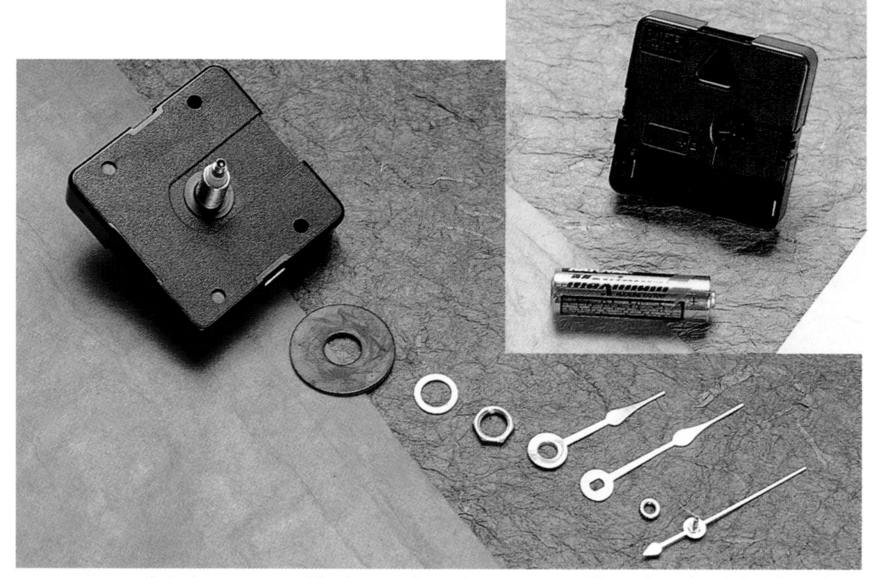

Battery-powered clock movements, like the one shown here, are typically packaged with hands and the washers and nuts necessary to attach them.

THE DIAL is the device, often featuring numbers, positioned on the face of the clock. You can purchase ready-printed dials that are usually disk shaped and feature precut holes that fit easily over the shaft. You can also create your own dial with everything from stamped-on numbers to fastened-on found objects. The template on page 122 will guide you in properly positioning numbers or other hour markers on a variety of sizes of round dials. For some clocks—especially those with a spare, modern look, you may decide to omit the dial from your design entirely.

Purchased dials, usually disk shaped, come in a variety of sizes and styles. You can also create your own dials by attaching number markers to your clock face. Here, designers used tiny vinyl surface guards (upper right) and colored gemstones (lower right). You can even omit the dial altogether, as the designer of this tin tile clock (upper left) did.

THE HANDS, often packaged with movements and also sold separately, are available in different shapes, sizes, colors, and styles. They're typically made out of plastic or metal.

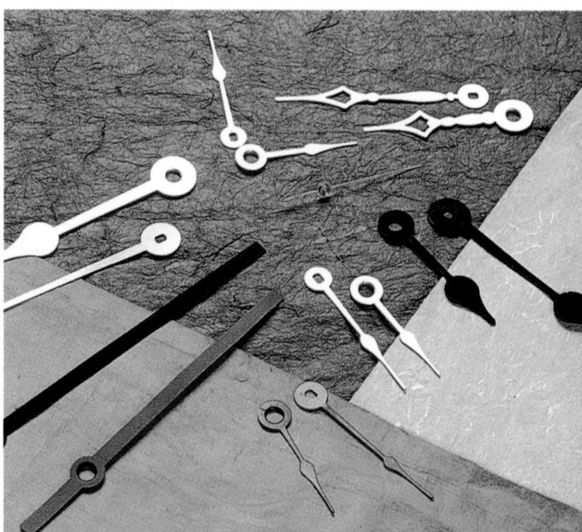

An assortment of purchased clock hands. Use them as they are, or embellish them.

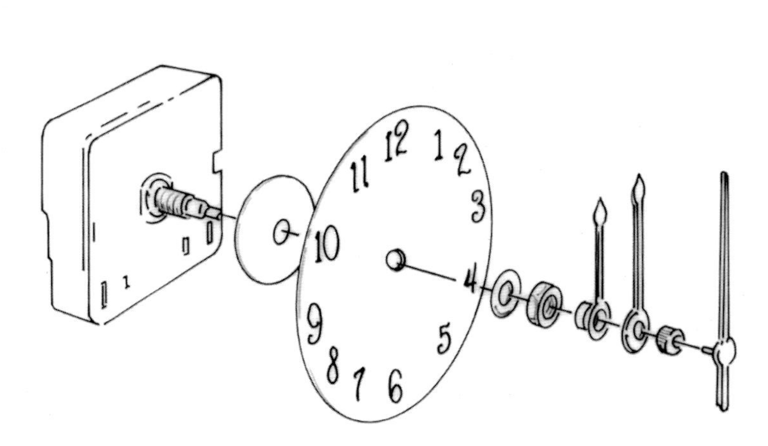

Attaching a dial and hands to a clock movement

Embellish them with paint, glitter, or instant rust products, if you like. Or, completely customize your clock hands by gluing accents on top of them (as designers did with the tiny shovel heads and twigs in the photos shown here). You can also drill holes in any number of objects and stick them on the movement shaft, in place of standard hands.

CLOCK BASES come precut and often predrilled, in lots of shapes, sizes, and styles, from simple circles to elaborate stands equipped to accommodate pendulum movements. If you want to get right to the painting and decorating, these solid-wood bases, available at craft stores, are the way to go. A number of projects in this book show you how to make the most of ready-made bases. Others show you how to start from scratch and create your own bases out of wood, metal, an imaginative array of found objects, even paper.

A BEZEL MOVEMENT is an alternative to the assemble-it-yourself clock face. It's a self-contained unit

Ready-made clock bases are available at craft stores. They come with predrilled holes for the movement shaft or with precut larger holes for bezel movements.

A purchased clock base, precut for a large bezel movement, like the one on the left. Two smaller bezel movements are also shown.

consisting of the clockworks, a metal frame, a dial, hands, and a glass cover. You insert the entire piece in a hole cut to size in the clock base.

PENDULUM MOVEMENTS are just like standard movements, except they're equipped with a pendulum shaft. Swinging pendulums attached to modern, battery-powered quartz clocks aren't responsible for moving gears (as they are on older mechanical clocks). They simply add a decorative touch.

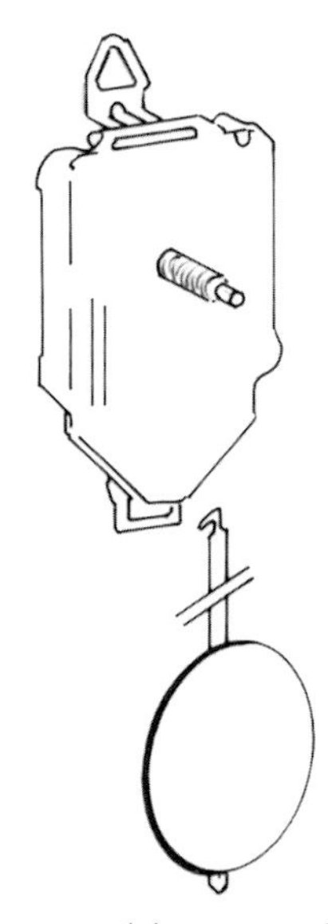

Attaching a pendulum to a pendulum movement

TIPS ON COMMONLY USED MATERIALS AND TECHNIQUES

DRILLING

Most of the projects in this book require you to drill a hole though whatever material your clock face is

Drilling a hole for a movement shaft

made of, so you can insert the shaft of the movement. A standard electric drill with a ⅜-inch (9 mm) chuck capacity will do the job in most cases. Occasionally, a project calls for drilling through glass or a mirror. If you don't have the tools or expertise for cutting that kind of hole yourself, simply take your piece to a glass supplier (the kind that sells tabletops, custom mirrors, etc.); they'll typically do the drilling for you for a nominal fee.

WOODWORKING

A few of the clock projects we feature involve some basic woodworking skills and equipment. Several call for a jigsaw (also called a saber saw), a hand-held power tool that's popular with home crafters for its ability to easily cut curves, free-form shapes, and large holes in boards up to 1½ inches (3.8 cm) thick. A shoe surrounding the blade can be tilted, so you can also use your jigsaw to make angled cuts.

Cutting curves with a jigsaw

POLYMER CLAY

This popular, easy-to-use, man-made material is the main ingredient in several clocks in this book. It's sold in craft stores in small blocks or in bulk quantities, and comes in a

wide a range of bright, contemporary colors. Polymer clay is a combination of polyvinyl chloride (PVC), plasticizer, and color pigments. It's as pliable as ceramic clay in its initial state. Once you've molded and sculpted it into shape, you simply "fire" it in a standard oven to harden it.

Even the softest brands of polymer clay are a bit stiff right out of the package, so you'll want to condition yours before working with it. Starting with a small amount, warm the clay in your hand or wrap it in wax paper and then a towel and place it on a sunny windowsill. Once it's pliable, roll the clay into a ball, then into a log. Stretch out the log, then fold it back on itself. Continue twisting and rolling the clay until it's an even consistency and color. Once it's conditioned, you can roll your polymer clay into a ball, wrap it in plastic, and store it in a cool, dry place for up to a week before using it.

Polymer clay is certified nontoxic, but it's a good idea to take a few precautions to minimize exposure to its chemical ingredients.

- Don't use kitchen utensils to prepare food once you've used them to manipulate polymer clay.
- Bake your polymer clay objects in a well-ventilated area, and clean the inside surfaces of your oven thoroughly with baking soda and water afterward.
- Wear rubber gloves while working with the clay, or wash your hands well afterward.

Transferring Patterns

At the back of this book are patterns for all of the roses, rabbits, and other flora, fauna, and shapes that show up on the clock projects we feature. Simply photocopy what you'd like to use, reducing or enlarging the pattern to fit the size of your project. Then transfer it in one of two ways.

1. Carefully cut around the edges of your copied pattern, tape or hold the silhouette to the surface you're working with, trace around the pattern with a pencil, then remove it and use the tracing as your guide when you cut, paint, or carve.
2. If you're transferring a pattern that's more intricate than the outline of a shape, use carbon paper. Put the carbon paper (carbon side down) on top of the surface you're transferring to (a piece of wood, for example), then place the copied pattern on top of the carbon paper. Draw over the lines of the pattern with a ballpoint pen. When you remove the pattern and the carbon paper, your design will appear in carbon on the wood.

Positioning Numbers

You need the numbers on your clock face to be perfectly placed (if you want the clock to keep accurate time, that is). But you'd rather leave all the tedious measuring to someone else. We've just made your day: flip to the Positioning Guide on page 122. It's a series of concentric circles, with points at 12 equally placed intervals to guide you in marking your hours. Reduce or enlarge the guide until one of the circles fits the clock face you have planned. Then use it to mark the points on the face where the numbers need to go.

Projects

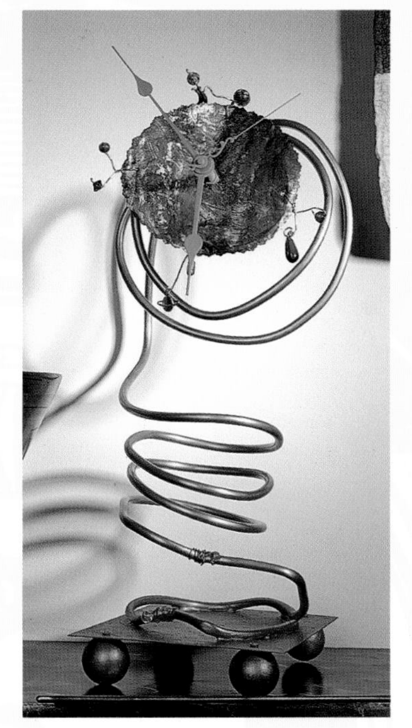

Scrap Metal Collage Clock

This unlikely assemblage of rusted metal, a cast-off sink stopper, and a photocopied face is urban sculpture and working clock in one.

Designer: Jean Tomaso Moore

"My inspiration for this clock came from a flattened and rusty metal scrap that I had found on the street several years ago. It reminded me of a piece of abstract art."

-Jean Tomaso Moore

MATERIALS

- Block of wood, 1 x 3 x 5 inches (2.5 x 7.5 x 12.5 cm)
- Thin wooden dowel (or metal rod) approximately 4 inches (10 cm) long
- Piece of rusty metal (or other found object)
- Water-based primer/sealer
- Spray paint (bronze metallic)
- Aluminum sink strainer/stopper
- Several inches (cm) of 24-gauge copper wire
- Face image, such as an antique moon or sun (You can clip your image from a magazine, copy it from a book or a card, etc.)
- Piece of rolled aluminum flashing approximately 12 inches (30 cm) square
- Acrylic craft paint in black, copper, bronze, and turquoise
- Double-sided foam mounting dots
- 12 tiny colored gemstones
- Cherub ornament (optional)
- Clock movement and hands

TOOLS

- Electric drill
- Artist's brushes
- Wire cutters
- Industrial-strength clear adhesive
- Double-sided tape
- White craft glue
- Foam brush (for applying the craft glue)
- Tinner's snips
- Work gloves (optional)

INSTRUCTIONS

1 Drill a hole in the center of the wooden block to accommodate the wooden dowel. Glue the dowel in place to serve as a support for the found object.

2 Coat the base with the primer. When it's dry, spray both the base and the dowel with bronze paint.

3 Remove the rubber stopper from the sink strainer, prime it, and paint it bronze as well.

4 Prop up the found object against the dowel, and thread the thin copper wire through the rust holes (or other openings) to attach it to the base. Add a dab of industrial-strength glue to secure the piece to the base.

5 Insert the clock shaft through the back side of the sink strainer, and use two pieces of double-sided tape to attach the movement to the back of the strainer.

6 Locate a spot on your found object where the clock face (sink strainer) will naturally sit. In the project shown, the designer used a jagged point of metal to "hang" the strainer. Secure the strainer in place wherever possible with more of the copper wire.

7 To make your face image sturdier, glue it to a small piece of aluminum flashing with white craft glue.

8 Use the tinner's snips to cut the backed image into a rounded shape. Then, apply watered-down layers of black, turquoise, copper, and bronze paint to simulate patina (the green film that forms naturally on exposed, aged copper and bronze).

9 Cut a circle shape about twice as large as the face image from the remaining aluminum flashing. Give it a distressed, aged appearance with the same painting process you used in step 8.

10 With double-sided foam mounting dots, create a three-dimensional effect of the face image hovering behind the found object and the larger circle floating behind the face. Stack the mounting dots together to create the level of depth you want at each layer, and attach the pieces.

11 Glue the tiny colored gems to the clock face to mark the hours.

12 Paint the hands turquoise. When they're dry, attach them to the shaft.

13 If you like, attach a cherub ornament or another embellishment to the found object with wire or glue.

Station Clock

REMINISCENT OF THE CLOCKS that kept train travelers racing through the station on schedule, this industrial-chic piece can stand on a mantel or be side-mounted on a wall.

DESIGNERS: ALLISON AND TRACY PAGE STILWELL

MATERIALS

- Round clock face, 11 inches (27.5 cm) in diameter
- Wooden base, 1-inch-thick (2.5 cm) circle, 6 inches (15 cm) in diameter
- Wooden base shaft, 2 x 2 x 5 inches (5 x 5 x 12.5 cm)
- White acrylic primer
- Acrylic paints in white, silver-gray, and black
- 12 lowercase "L" press-type letters (optional) (Press type is available at stores that sell craft, art, and office supplies. It takes the stress out of printing and painting neat, consistent letters and numbers. Be sure to add several coats of sealer to protect press-type symbols.)
- Acrylic sealer
- Wood screw, 1½ to 2 inches (3.8 to 5 cm), to attach the shaft to the base
- Sheet of thick vellum or of the plastic material sold at craft stores for making stencils
- Black permanent marker
- Clock movement and hands

TOOLS

- Ruler
- Pencil
- Drill with a ½-inch (1.3 cm) bit and a countersink bit
- Artist's brushes, fine tipped (optional) and medium
- Medium-grade sandpaper
- Wood glue
- Book or heavy object (optional)
- Scissors or craft knife
- Craft glue

INSTRUCTIONS

1 If you are not using a clock face with a pre-drilled hole, use the ruler to find the center of the clock face and mark the point. Drill a hole in the center of the face for the clock movement.

2 Using an artist's brush, apply one coat of the white acrylic primer to all of the wooden surfaces. Let the primer dry completely, then use the sandpaper to sand the parts as necessary.

3 Apply several coats of paint to the clock parts. Paint the clock base, the shaft, and the edge of the clock face silver-gray and paint the clock face white. Allow the paint to dry completely between coats.

4 With the pencil, lightly mark the locations of number lines around the clock face. You can use the lowercase version of the letter "L" from a simple style of press type for your number markings (an easy way to achieve consistent, clean lines). Or, with the fine-tipped artist's brush and black paint, draw ¾-inch-long (1.9 cm) lines around the face. After applying or painting the symbols, use a brush to apply several coats of acrylic varnish or sealer to protect the clock face.

5 With the ruler, find the center of the bottom of the clock base, mark the spot, and predrill a hole.

6 Apply a coat of wood glue to the bottom of the shaft (around the hole) and to the top of the clock base to help reinforce the assembly, then screw the shaft to the base, countersinking the screw.

7 Measure 2½ inches (6.3 cm) down from the top of the wooden shaft, and make a mark. (You will glue this portion of the shaft to the back of the clock face.) Apply a coat of wood glue to this section of the shaft and to the back of the clock face, where it will meet the shaft (directly behind the 6 o'clock location on the front of the clock face). Allow the glue to set up for several moments, then press the pieces firmly together and let them dry overnight. You may want to set a book or another heavy object on top of the shaft to ensure a strong bond.

8 With the marker, draw two simple, thick arrows on the vellum or stencil paper. You'll glue these pieces onto your clock hands in the next step; make them small enough so that the clock hands will still be able to work properly and fit on your clock face. Cut out the arrows with the scissors or craft knife, then use an artist's brush to apply one or two coats of the black acrylic paint, covering the shapes completely.

9 Once the arrows are dry, glue them to the ends of the purchased clock hands. Allow the hands to dry completely.

10 Insert the clock movement and attach the hands.

VARIATION

You can use almost any symbol to signify numbers on the face—dots, lines, spirals, squiggles—and every hour could be different.

Playtime

A STANDARD APPLIANCE TIMER (like you might use to set a lamp so it clicks on after dusk) makes this ingenious seesaw clock work—and play. At noon and midnight the teetering ants are level. In between they totter up and down. What better way to track the hours 'till the end of the workday.

DESIGNER: DON SHULL

MATERIALS

- Strip of lightweight wood $\frac{1}{4}$ x $\frac{7}{8}$ x 12 inches (.6 x 2.2 x 30 cm)
- $\frac{7}{8}$ inch (2.2 cm) of brass tubing $\frac{1}{8}$ inch (3 mm) in diameter
- $1\frac{1}{2}$ inches (3.8 cm) of dowel rod $\frac{1}{4}$ inch (6 mm) in diameter
- $1\frac{1}{2}$ inches (3.8 cm) of dowel rod $\frac{1}{8}$ inch (3 mm) in diameter
- Acrylic paint in color(s) of your choice
- Small block of wood approximately $\frac{5}{8}$ inch (1.6 cm) thick, cut to a length and width that matches the top of the timer
- Appliance or lamp timer (available at hardware stores)
- 2 No. 6 x $\frac{1}{2}$-inch (1.3 cm) wood screws
- 12 inches (30 cm) of 12- or 14-gauge galvanized wire
- 6 inches (15 cm) of dowel rod $\frac{3}{8}$ inch (.9 mm) in diameter
- 24 inches (61 cm) of 18-gauge galvanized wire
- 6 inches (15 cm) of 24-gauge galvanized wire
- Black spray paint

TOOLS

- Ruler
- Pencil
- Saw
- Electric drill with drill bits $\frac{1}{16}$, $\frac{1}{8}$, and $\frac{1}{4}$ inch (1.5, 3, and 6 mm) in diameter
- Wood glue
- Artist's brush
- Wire cutters
- Coarse rasp or file

INSTRUCTIONS

1 Locate the centerline of the 12-inch (30 cm) strip of wood and mark it. Using the mark as a guide, drill a cross hole $\frac{1}{8}$ inch (3 mm) in diameter through the width of the piece. Push the piece of brass tubing into the hole to create a bearing for the seesaw.

2 Mark centered points $2\frac{1}{2}$ inches (6.25 cm) in from each end of the strip, and drill holes $\frac{1}{4}$ inch (6 mm) in diameter for the handle bases.

3 Mark a point $4\frac{1}{2}$ inches (11.3 cm) in from the right end of the strip, and start a hole $\frac{1}{16}$ inch (1.5 mm) in diameter with the drill. This hole will hold one end of the wire that moves the seesaw. See figure 1 for placement of holes for steps 1 through 3.

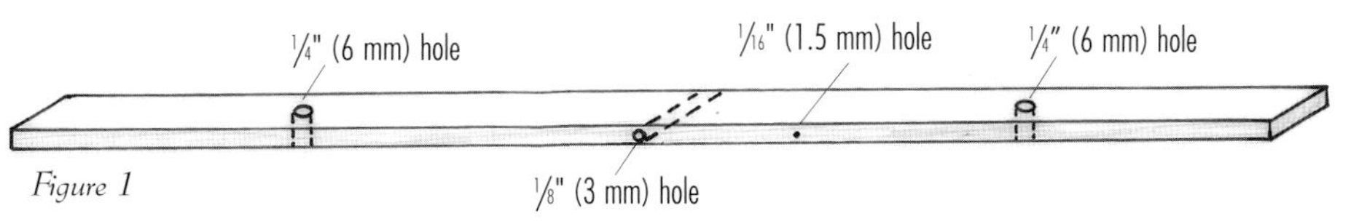

Figure 1

4 Cut the $\frac{1}{4}$-inch (6 mm) dowel rod into two $\frac{3}{4}$-inch (1.9 cm) pieces. On each, measure $\frac{1}{4}$ inch (6 mm) down from the top, and drill a hole $\frac{1}{8}$ inch (3 mm) in diameter all the way through. See figure 2.

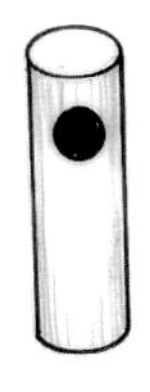

Figure 2

5 Cut the $\frac{1}{8}$-inch (3 mm) dowel rod into two $\frac{3}{4}$-inch (1.9 cm) pieces. Stick the smaller dowel rods through the holes in the larger dowel rods to form handles, and glue the handle bases into the holes you drilled in step 2. See figure 3.

Figure 3

6 Paint the seesaw seat and handles, and let the piece dry.

7 Use the saw to angle the block of wood so it forms a peak, which will provide clearance for the seesaw when the block sits atop the timer.

8 Carefully take apart the timer so that you can use the wood screws to mount the angled block on top of it, screwing up through

the top of one half of the timer. Put the other half of the timer back in place. Paint the unit, if you like.

9 Cut a 2¼-inch (5.6 cm) piece of the 12- or 14-gauge wire, push it through the hole with brass tubing in the center of the seesaw, and bend the ends down. Hold the seesaw over the angled block, and use the ends of the wire to mark where two holes should be drilled. Make the holes 1/16 inch (1.5 mm) in diameter. Fit the wire in the holes to position the seesaw on top of the timer unit. See figure 4.

10 Out of the remaining 12- or 14-gauge wire, bend the piece that will connect the seesaw to the timer dial and move the seesaw up and down. Make the first bend 1¼ inches (3.1 cm) from one end, push the wire into the hole on the seesaw, then move the seesaw up and down to determine where the hole on the timer should be. Drill the hole 1/16 inch (1.5 mm) in diameter, and bend the wire to fit. The seesaw should be able to move through a full 24-hour cycle without touching the block on top of the timer. See figure 5.

11 Cut two pieces measuring 1¾ inches (4.5 cm) from the dowel rod that is ⅜ inch (9 mm) in diameter. Each will become an ant body. Create each ant by putting the piece of dowel rod in the drill, turning it, and using a coarse rasp or file to shape the bodies. Use figure 6 as a guide.

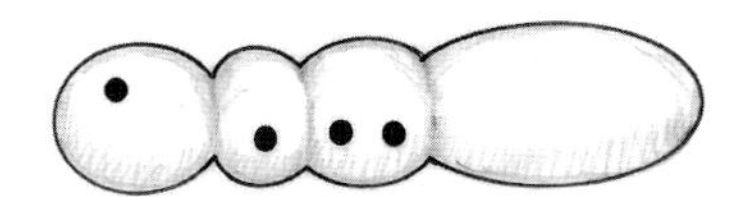

Figure 6

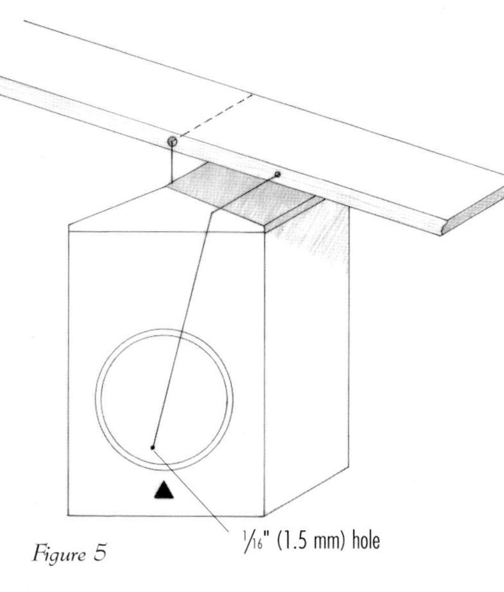

Figure 5

12 Following the marks in figure 6, start small holes with the drill on each side of each ant body. Use the 18-gauge wire to create legs and the 24-gauge wire to make antennae. Insert each wire piece in a starter hole, using the wire to open up the hole as much as necessary.

13 Spray paint the finished ants black. Once they're dry, attach them to their seats on the seesaw.

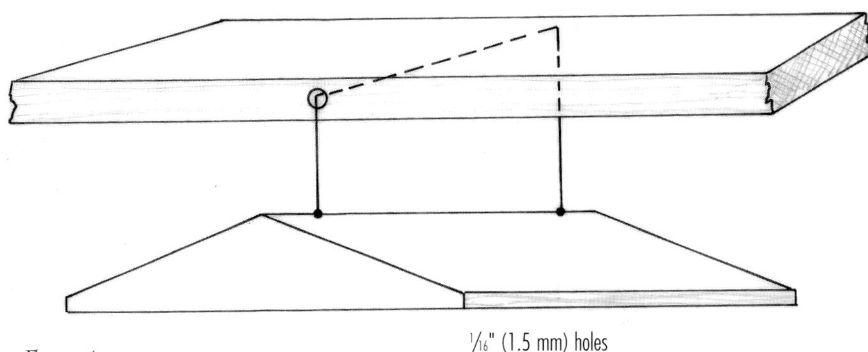

Figure 4

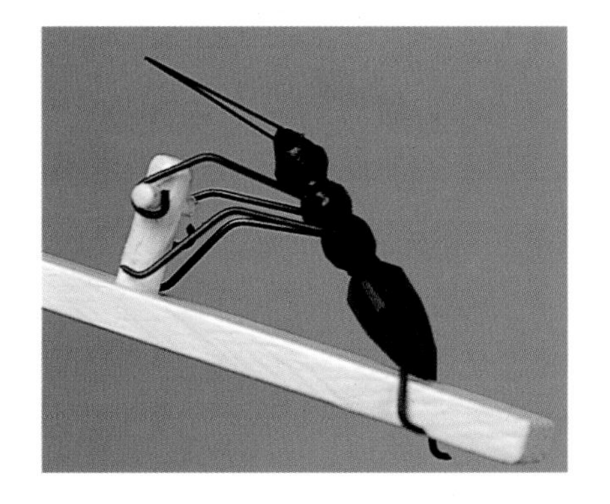

Island Time

The traditional Hawaiian greeting, "Aloha," has welcomed newcomers to Honolulu Harbor since 1926—in a big way. It's carved into the white cement of the grand Aloha Tower, a 184-foot (55 m) navigational structure adorned with four giant clock faces. During the "Boat Days" of the 1930s, the Aloha Tower stood as a centerpiece for festive celebrations staged for luxury steamships and their celebrity passengers, who were met by ambassadors including hula dancers and the Royal Hawaiian Band.

The clocks themselves match the grandiosity of the tower. Each of the clock faces, which are positioned around the tower in the four directions of the compass, measures 12½ feet (3.8 m) in diameter. The movements of each weigh 7 tons (6.3 metric tons), and the balance wheels, which regulate the movements, travel 3,186 miles (5,097 km) a year. Boats drifting as far as 5 miles (8 km) out at sea can use the illuminated faces to check the local time or confirm their direction according to the compass points on the clocks' faces. Sirens have also been used to announce the clock's time, and boat captains once set their watches to noon each day when a 4-foot-wide (1.2 m) ball was dropped from the mast atop the tower.

Through the decades, the Aloha Tower has served as housing for customs offices, arrival and departure space for boat passengers, home for harbor pilots and the harbor master, a control facility for convoy shipping during World War II, and, of course, as master timekeeper for the island.

Skyscrapers now overshadow the tower, and most visitors to Honolulu (celebrity and otherwise) arrive by airplane rather than steamship. Still, the Aloha Tower holds fast to its place in history and continues to tick away—157,680,000 ticks a year, to be exact.

Barn Wood Mantel Clock

DESIGNER: ROLF HOLMQUIST

WEATHERED WOOD, crackle glaze, and recycled ornaments, hinges, and handles make a handsome, sturdy clock that appears to have stood the test of time. Don't cross this project off your list if you don't have a table saw that's ready to whir into action. Get a friend who knows all about angle cuts and grooves to prepare the wood according to the lists we've provided (nothing out of the ordinary here; it's all basic). Then, simply enjoy the fun of putting the pieces together.

CUT LIST

Desc.	Qty	Material	Dimensions
Clock front	1	Barn wood	1 x 28 x 13½ inches (2.5 x 71 x 34.5 cm)
Clock back	1	Barn wood	1 x 28 x 13½ inches (2.5 x 72 x 34 cm)
Clock sides	2	Barn wood	1 x 21¼ x 5 inches (2.5 x 53.8 x 12.5 cm)
Clock base	1	Standard pine	2 x 17½ x 9 inches (5 x 43.8 x 22.5 cm)
Foot blocks	4	Standard pine	4¾ x 2¾ x 2¾ inches (11.9 x 7 x 7 cm)
Cleat	1	Standard pine	¾ x 1 x 9 inches (1.9 x 2.5 x 22.5 cm) (to attach back to base)
Cleats	2	Standard pine	¾ x 1 x 4 inches (1.9 x 2.5 x 10 cm) (to attach sides to base)
Cleats*	2	Standard pine	¾ x ¾ x 10 inches (1.9 x 1.9 x 25 cm) (for holding mirror)
Strips	4	Standard pine	¾ x ½ x 3 inches (1.9 x 1.3 x 7.5 cm) (to attach glass to front)
Right roof piece**	1	Standard pine	¾ x 8-¾ x 9 inches (1.9 x 22.5 x 23 cm)
Left roof piece**	1	Standard pine	¾ x 8¼ x 9 inches (1.9 x 20.7 x 22.5 cm)
Roof bases**	2	Standard pine	¾ x 8 x 4 inches (1.9 x 20 x 10 cm)
Cupola pieces	2	Standard pine	¾ x 2 x 1 inches (1.9 x 5 x 2.5 cm)
Roof triangle insert	1	Standard pine	¾ x 2 x 2 x 2 inches (1.9 x 5 x 5 x 5 cm)

*Score cut the center of each of these cleats to the thickness of the mirror.
**The right, left, and base roof pieces have angle cuts (see step 15). These measurements are after angles have been cut.

MATERIALS

- Stain (optional)
- 2 old hinges
- 5 old kitchen drawer handles
- Varnish
- 4 wooden doll heads
- Latex paint (flat) (You'll want a dark color for the base coat and a light color for the top coat.)
- Crackle glaze
- Mirror, 12 inches (30 cm) square
- Piece of single-strength glass, 7 inches (17.5 cm) square
- Wooden curtain-rod end
- 2 round wooden balls, 1¼ inches (3.1 cm) in diameter
- 2 round wooden balls, ¾ inch (1.9 cm) in diameter
- Clock face, 6 inches (15 cm) in diameter (Recycle one, if you like.)
- Frame and glass covering from an old kitchen clock
- Approximately 48 1½-inch (3.8 cm) drywall screws
- Approximately 10 1-inch (2.5 cm) drywall screws
- Found decorative piece for front of roof (optional)
- Pendulum clock movement and hands (Most barn wood is 1 inch [2.5 cm] thick. Since your movement shaft will be inserted through barn wood, purchase one designed to fit thicker clock faces. If you're using a standard-size movement, which typically fits a ¾-inch [1.9 cm] face, we'll tell you how to chisel the barn wood to make it fit in step 21.)

TOOLS

- Table saw (for precise cuts)
- Jigsaw (for curves and angle cuts)
- Drill with a 1/8-inch (3 mm) drill bit, a 1/4-inch (6 mm) drill bit, and a countersink bit
- Sandpaper
- Stain brush
- Tape measure
- Pencil
- Paintbrush
- Glass cutter (optional)
- Glue
- Wood chisel (optional)
- Wood filler

Back view of clock

INSTRUCTIONS

1 Cut the front piece, following the illustration in figure 1. For the 5-inch (13 cm) square on the front (the pendulum window), first drill out each corner with the 1/4-inch (1 cm) bit, so you end up with rounded corners. You can then insert the jigsaw for your cuts. Sand the edges of the square opening, and stain them if you like. Approximately 6 1/2 inches (16.3 cm) above the opening, drill the hole for the clock movement.

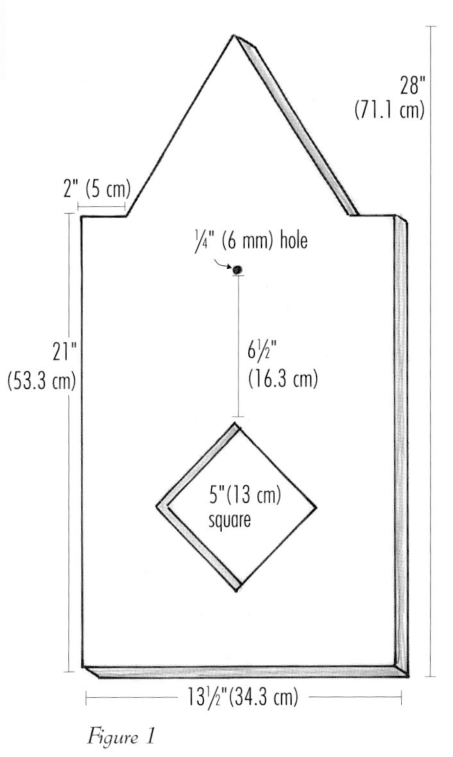

Figure 1

2 Cut the back piece, following the illustration in figure 2. Lay the front piece on top of the back piece. Through the drilled hole on the front piece, mark the corresponding spot on the back piece. Position your access door around this hole. (The access door will allow you to reach the clock movement and pendulum once your clock is assembled.)

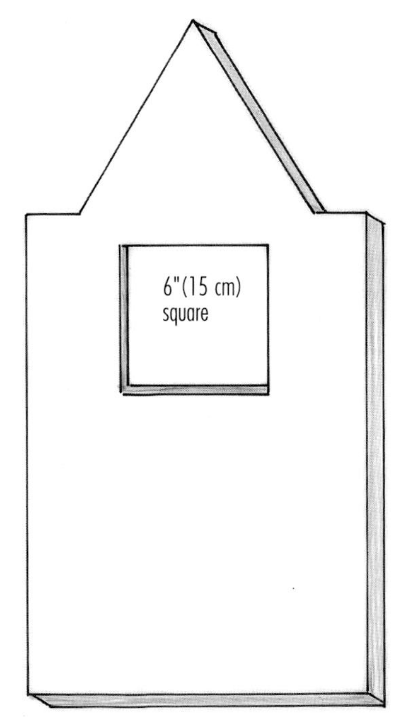

Figure 2

3 Hinge the access door and attach one of the kitchen drawer handles.

4 Varnish the front, back, and two side pieces of barn wood.

5 Bevel cut the base piece by angling the saw at 45°.

6 Drill a hole in the center of each foot block. (You can easily find the center by connecting each corner with a pencil line.) Screw a wooden doll head to each

the mirror along these lines. Stain the cleats, including the slots where the mirror will be held.

12 Cut the mirror piece (or have it cut at a glass supplier) so it will slide snugly into the slotted holders.

13 Attach the square of glass to the inside of the square opening on the front piece by screwing the four holding strips around the edges of the glass (see figure 5).

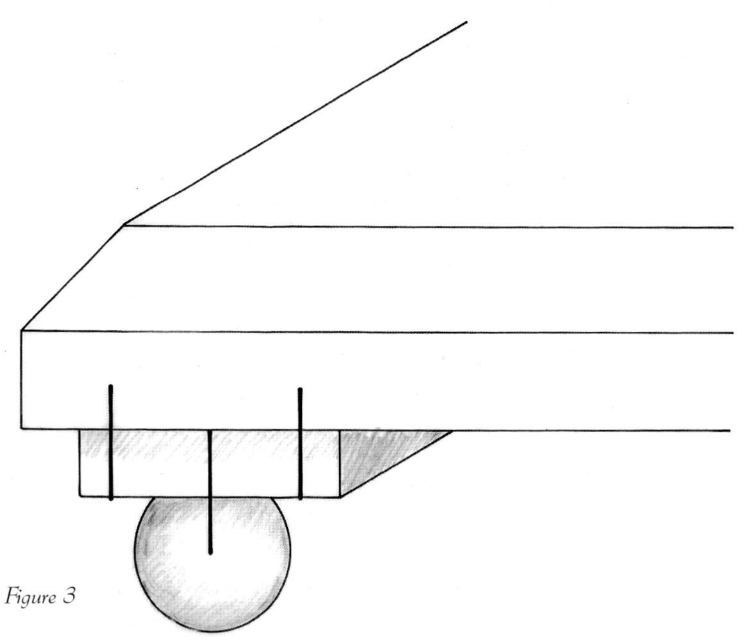

Figure 3

block, turn the block over, and screw an assembled foot to each corner of the base (see figure 3). When attaching the feet to the base, pre drill holes for the screws, and position the feet just slightly in from the corners, rather than flush with the edges of the base.

7 Paint the completed base with a dark base coat, a lighter top coat, and a layer of the crackle glaze. Follow the manufacturer's directions for applying the glaze.

8 Screw the two side pieces to the back piece, positioning the side pieces approximately ¾ inch (1.9 cm) in from the edges of the back piece (see figure 4).

9 Attach the cleats to the back and side pieces on the inside of the clock. Predrill the holes, then

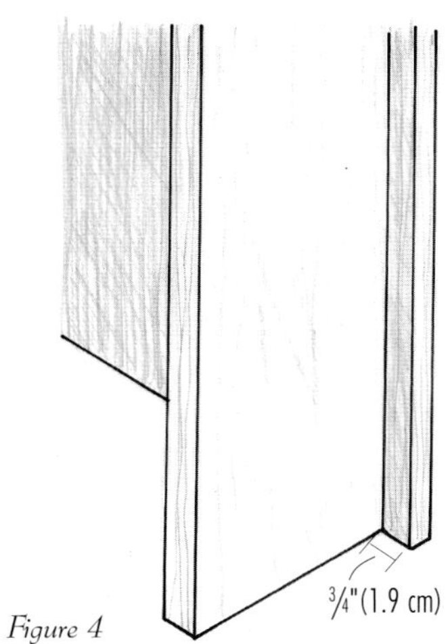

Figure 4

screw the pieces together from the outside.

10 Attach the entire piece to the base, predrilling holes and inserting the screws from the cleats to the base.

11 On the inside of each side piece, mark the center, then draw a vertical line down to the base. Screw the cleats for holding

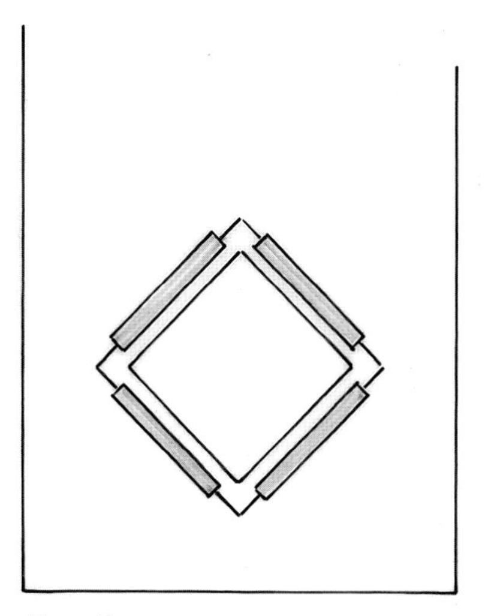

Figure 5

14 Screw the four remaining kitchen drawer handles to the sides of the clock for decorative effect. The designer of this project positioned his handles 2 inches (5 cm) up from the bottom and 2 inches (5 cm) down from the top.

15 Temporarily position your right and left roof pieces and two roof base pieces, following the illustration in figure 6. Make angle cuts as necessary to fit the pieces together.

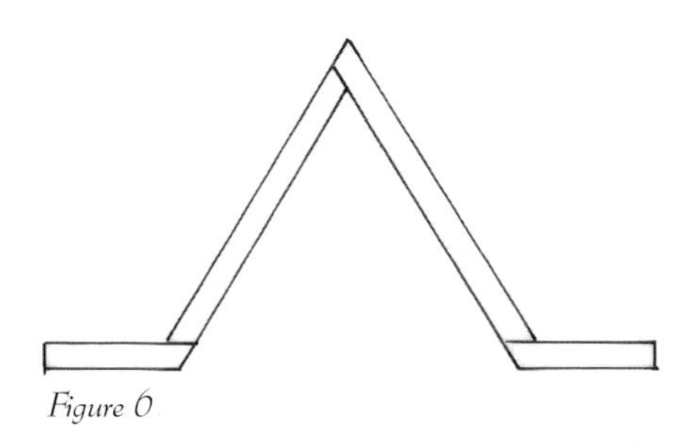

Figure 6

16 Predrill holes and screw the right and left roof pieces together.

17 Mark the middle of the peak, and glue the two triangular cupola pieces in place on top of the mark. Once the glue is dry, screw the curtain-rod end on top of the cupola.

18 Glue the roof triangle insert under the peak of the roof, flush with the front (see figure 7).

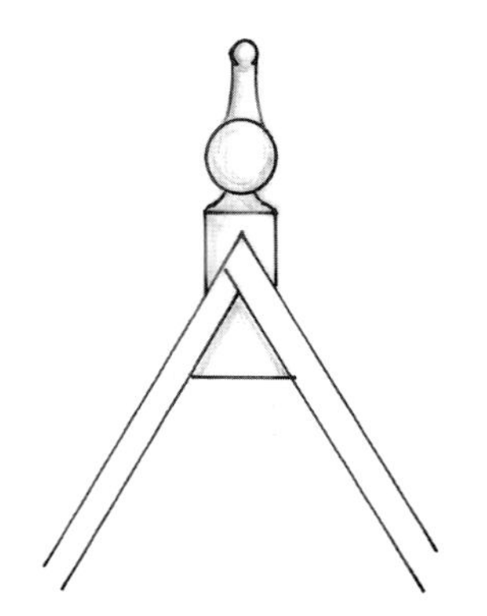

Figure 7

19 Glue the two small wooden balls on top of the two large wooden balls. When the glue is dry, screw one to each roof base piece, approximately ⅝ inch (1.6 cm) in from the front edge and 1½ inches (3.8 cm) in from the side edge. Use the project photo as a guide for placement.

20 Before assembling the remaining pieces of the clock, paint and glaze all of the roof pieces and the kitchen-clock frame to match the base.

21 If your movement shaft isn't long enough to fit through the barn wood, chisel out some space on the inside of the front piece (around the hole you drilled), until the shaft fits.

22 Attach the front piece to the rest of the clock by predrilling holes and using six 1½-inch (3.8 cm) screws to connect the edges of the front and side pieces.

23 Insert the clock movement, add the face, and attach the hands.

24 Attach the recycled kitchen-clock frame and glass with two 1-inch (2.5 cm) screws.

25 Attach all the roof pieces by predrilling holes, countersinking the screws, and finishing with wood filler and sandpaper.

26 If you like, glue a found decorative piece to the front of the roof.

Can Someone Tell Me the Bloomin' Time?

Niagara Falls, Ontario, Canada. It's home to one of the world's more famous natural wonders. But it's also the site of one of history's larger horological—and horticultural—accomplishments: a giant floral clock.

Niagara Falls' landmark clock was inspired by Scotland's well-known floral clock, built in 1903 in Edinburgh's Princess Street Gardens. The former chairman of Canadian Hydro proposed the Canadian version, then had his company's employees construct it. They completed the clock, a full 40 feet (12.2 m) in diameter, in 1950.

More than 25,000 plants are carefully placed in the soil to create a new clock face each year. Standard varieties include colorful violas and crocus in the spring. Later in the year, they're replaced with red and yellow Joseph's coat, purple-gray cotton lavender, and traditional green bedding plants. The flowers and plants are tended weekly and displayed year-round. If there's a chance of snowfall, the second hand, which measures 21 feet (6.5 m) long, is removed to protect it from damage. The hour and minute hands (not quite as long) typically weather the storms. (All of the hands, by the way, are made of stainless steel tubing and have a combined weight of 1,250 pounds [567 kg].)

Thanks to the Scots, the idea of combining hours and flowers is popular worldwide. Their floral clock in Edinburgh, considered the oldest, inspired not only the Canadians, but many others. There's a floral clock in Toi Town, Izu Province, Japan (at 102 feet [31 m] in diameter, it's also listed in the Guiness Book of World Records as the world's largest clock). Veracruz, Mexico, lays claim to the world's only double-faced floral clock, which also plays 10 different melodies, including the recognizable Westminster Chimes. And floral clocks bloom in Tasmania and Melbourne, Australia; Geneva, Switzerland; Oostende, Belgium; and in the tiny town of New Glarus, Wisconsin, USA.

Robot Clock

DESIGNERS: ALLISON AND TRACY PAGE STILWELL

THIS ENDEARING LITTLE automaton combines high-tech and heart. Part woodblock toy and part computer-age creation, he'll keep time happily in cubicles and playrooms alike.

CUT LIST*

Body	1	2 x 4, $4\frac{1}{4}$ inches long (5 x 10 x 11.9 cm)
Arms	2	1 x 1, $4\frac{3}{4}$ inches long (2.5 x 2.5 x 11.9 cm), tops cut to a 45° bevelled edge for "shoulders"
Legs	1	2 x 2 inches (5 x 5 cm)
Hands	2	$\frac{1}{2}$ x $\frac{1}{2}$ inch (1.3 x 1.3 cm) wooden spools
Feet	2	1 x 1 x 1 inches (2.5 x 2.5 x 2.5 cm)

*The labels used to describe standard-size lumber, such as 2 x 4 (inches) and 1 x 1 (inches), do not reflect exact measurements. The metric equivalents listed here are for the actual dimensions of the lumber, and therefore may not correspond exactly to the standard-size labels.

MATERIALS

- White acrylic primer
- Acrylic paints in assorted colors (Periwinkle blue, silver, silver-gray, white, and black were used in the project shown.)
- Black permanent marker, fine tipped (optional)
- Thin-gauge silver-colored wire, straight or coiled
- Small beads, with holes for threading
- Purchased wooden globe-shaped base for a bezel clock movement (If you can't find or order this base, cut a wooden circle 2⅛ inches [5.3 cm] in diameter, with a 1½ inch [3.8 cm] opening for the bezel insert.)
- 2 nails, 1½ inches (3.8 cm) long, with wide heads
- Mini bezel movement

TOOLS

- Artist's brushes, small and medium
- Medium-grade sandpaper
- Wood glue
- Ruler
- Pencil
- Wood carving tool or sharp craft knife
- Craft glue
- Drill with small drill bit
- Hammer

INSTRUCTIONS

1 With an artist's brush, apply one coat of the acrylic primer to all of the wooden surfaces. Let the pieces dry completely, then sand them as necessary with the sandpaper.

2 Glue the leg piece to the body with wood glue, applying the glue to both surfaces and allowing it to set up for several minutes before attaching the pieces. Allow the glue to dry completely before handling the pieces.

3 Paint the hands and feet silver-gray and paint the remaining body parts periwinkle blue. Let the paint dry completely, then use the sandpaper to sand the pieces as necessary.

4 With the ruler and pencil, lightly mark a line down the center of the front of the leg piece. Using the mark as a guide, carefully carve a line. Carve just deeply enough to take off the paint so the wood shows through, giving the impression of two separate legs.

5 Decorate the front of the robot with dials and gauges, painting them on the body in an assortment of colors such as silver, silver-gray, white, and black.

6 Attach wire gadgetry, as well. If your wire is straight, wrap it around the pencil or another slender, cylindrical object to create a coil. Then, insert each end of the wire coil into the hole of a bead, and secure the bead with a dab of craft glue. Once they're dry, glue the pieces (with the beads as sticking points) to the robot body, allowing the coiled wire to wrap around an edge or hang freely between the beads.

7 Create a few other interesting bead-and-wire assemblies and glue them to the wooden globe, which will be the head piece.

8 Use the wood glue to attach the hands (the spools) to the bottom of each arm piece, and the feet (1-inch [2.5 cm] square blocks) to the front bottom edge of the leg piece. Allow the glue to dry completely.

9 Drill a small hole in the "shoulder" of each arm, about ½-inch (1.3 cm) down and centered. Use the hammer and nails to attach the arms at the top center of each side of the body. Don't hammer the arms in too tightly; leave them loose enough to swing freely.

10 Insert the mini bezel movement in the clock face. Use the wood glue to glue the globe (or head piece) to the top center of the body, and allow it to dry completely.

Copper-Coil Clock

WHAT IS IT THAT'S SO ENGAGING about a strand of coiled copper tubing? Maybe it's that it looks like it's getting away with something. Shouldn't it be inside some appliance, making it run, instead of out in the open like this, bobbing away behind the face of a clock? What fun to break the rules!

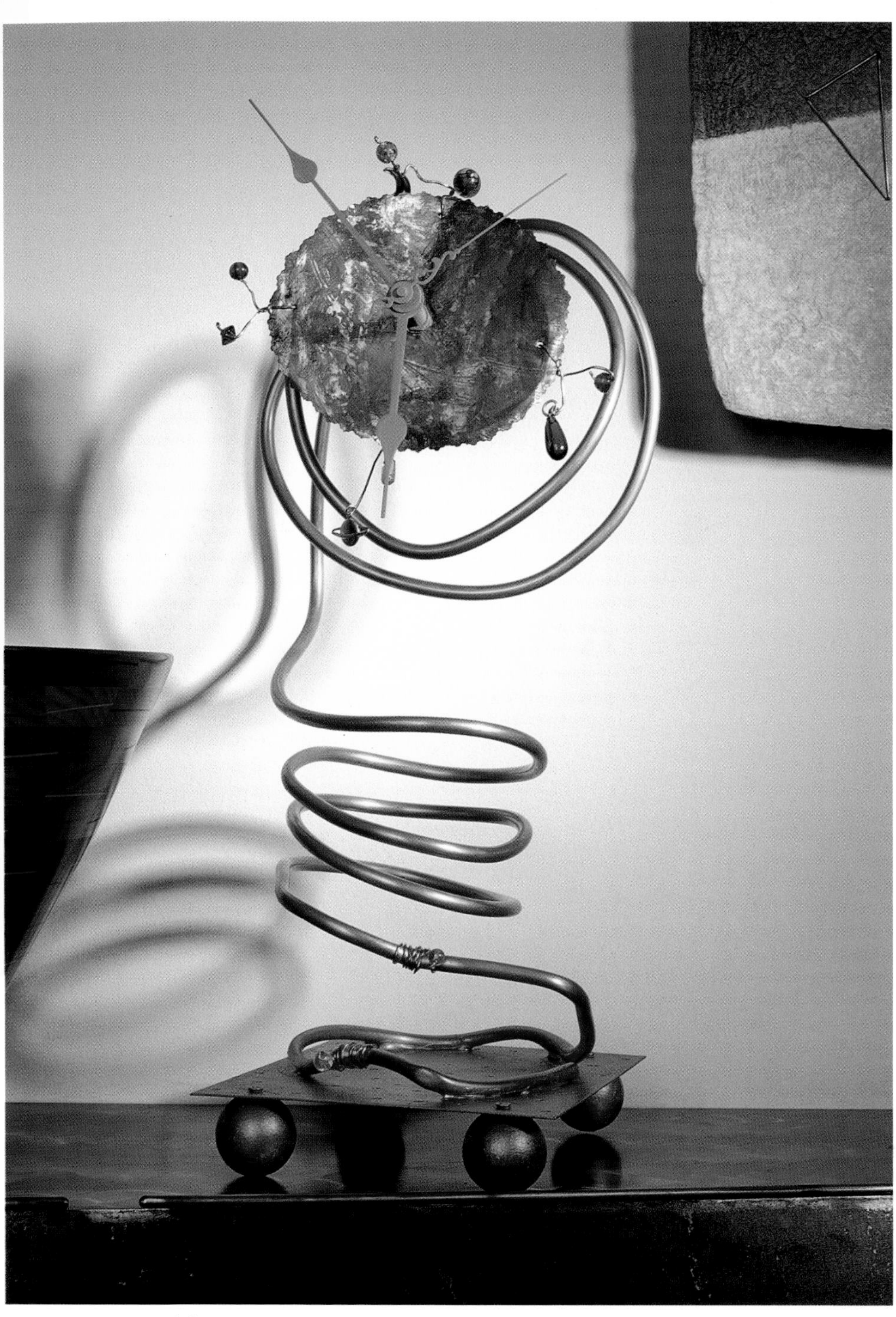

DESIGNER: JEAN TOMASO MOORE

MATERIALS

- 10 feet (3 m) of ¾-inch (1.9 cm) flexible copper tubing
- Piece of rolled aluminum flashing approximately 4 x 6 inches (10 x 15 cm)
- Acrylic craft paints in black, turquoise, bronze, lime green, orange, and metallic copper
- Water-based primer paint
- Metal tie plate, 5¾ inches (14.5 cm) square (Tie plates, available at home centers, are used as roof trusses.)
- 4 wooden balls 1¾ inches (4.5 cm) in diameter
- Spray acrylic fixative/sealer
- Several feet (meters) of 18-gauge copper wire
- 4 screws
- Several glass beads and charms
- Clock movement and hands

TOOLS

- Tinner's snips or deckle-edge craft scissors
- Nail
- Hammer
- Artist's brushes and/or painting sponges
- Double-sided tape
- Screwdriver
- Industrial-strength clear adhesive
- Bent-nose pliers
- Work gloves (optional)

INSTRUCTIONS

1 Twist the copper tubing into a spiral shape, creating a natural base with the bottom spiral. Bend the spiral forward at the top (the last three twists or so). The clock face will attach to the front spiral of this top section. Use the project photo as a guide when shaping your spiral.

2 With the tinner's snips or the deckle-edge scissors, cut the piece of aluminum flashing into an oval shape for the clock face. The face in the project shown measures 3½ inches (8.8 cm) across on its short side and 4½ inches (11.3 cm) across on its long side. The designer chose to cut it small enough so that the hands would extend past its edges.

3 Using the point of the snips, poke a hole in the center of the clock face. Then, use the nail to poke four holes around the perimeter of the face (each about ½ inch [1.3 cm] in from the edge) at the 12, 3, 6, and 9 o'clock positions.

4 Hammer down any jagged edges that were created by poking the holes, then use the hammer or any other blunt tool to texturize and distress the clock face.

5 On the clock face, brush or sponge on watered-down layers of black, turquoise, bronze, and metallic copper paint to simulate patina (the green film that forms naturally on exposed, aged copper and bronze).

6 Prime the clock hands and let them dry. Paint each in a different bright color (turquoise, orange, and lime green were used in the project shown) to make them stand out against the face.

7 Prime the metal tie plate (which will serve as the clock base), and paint it metallic copper. Paint the four wooden balls metallic copper as well.

8 Once all the painted pieces are dry, spray them with a coat of acrylic sealer.

9 To attach the clock face to the spiral of copper tubing, thread 3 or 4 inches (7.5 to 10 cm) of 18-gauge wire through each of the nail holes on the face and around the front loop of the spiral. Twist the ends to hold the face in place, leaving "tails" of wire sticking out.

10 Use two strips of double-sided tape to attach the clock movement to the back of the face.

11 Screw a wooden ball onto each corner of the tie plate to create four feet on the clock base.

12 Center the copper tubing spiral on the tie plate. Use industrial-strength adhesive to attach the tubing to the base at points of contact. After the adhesive dries, paint any adhesive that is visible metallic copper so it blends with the base.

13 Attach the clock hands, embellish the wire "tails" around the face with beads and charms, and wrap lengths of wire around the tubing, adding beads and charms to the ends. To hold the beads and charms in place, use the bent-nose pliers to make spiral loops on the ends of the wire.

London's Towering Timekeeper

Tower clocks first appeared in 14th-century Italy. Since then, they have become symbols of civic pride in communities around the world, displaying the hour from positions of authority atop town halls, churches, university unions, and other important structures. The model for many resides in St. Stephen's Tower in the Palace of Westminster. Its official name is the Westminster Clock. But most of us know it by its chummier alias: Big Ben.

Seven years of debate preceded the building of the Westminster Clock. Astronomers, politicians, clock makers, watchmakers, architects, and lawyers argued the ins and outs of constructing a clock that would live up to the precise standards required of a "king's clock." Finally, clock enthusiast Edmund Beckett Denison and a family of watchmakers named the Dents formed a partnership to tackle the job. They finished making the now-famous clock in 1859.

Gothic-numeral faces 22½ feet (6.85 m) in diameter adorn each side of the tower. They feature hands that are 14 feet (4.2 m) and 9 feet (2.7 m) long. The clock's mechanism, now run by an electric motor, was originally gravity based, requiring that two men spend 30 hours a week drawing up the mighty weights that ran the works.

Held to high standards, the Westminster Clock is a symbol of national pride. It's illuminated at night while the British Parliament is in session, and its four bells, which weigh between 1 and 4 tons (.9 to 3.6 metric tons) each, ring out over London daily, dependably announcing each hour.

Clock Tower Trivia: Big Ben, the nickname most of us use for the Westminster Clock, is actually the name of one of the bells that rings out the time from the top of the tower.

The Clock That Time Undid

ANY CHANCE THIS CLOCK looks like a metaphor for a typical and recurring time in your life? Monday morning, maybe? If so, schedule a Saturday afternoon therapy session to create your own fallen-apart clock.

MATERIALS

- Top from an old wooden ironing board or 3/4-inch (1.9 cm) plywood measuring approximately 54 inches (137 cm) high and 12 3/4 inches (31.9 cm) wide, cut so that it tapers at the top and resembles a clock tower
- 2 pieces of wood measuring approximately 1 1/2 x 1 1/2 x 8 inches (3.8 x 3.8 x 20 cm)
- Natural-wood block set
- Wooden balls in various sizes
- Wooden numbers from 1 to 12 (measuring approximately 2 inches [5 cm] high) with self-stick adhesive backs
- 2 neon-colored plastic Slinkys
- Piece of 1/8-inch (3 mm) Masonite measuring approximately 15 x 15 inches (37.5 x 37.5 cm)
- Gesso primer
- 7-inch (17.5 cm) embroidery hoop (You'll use the inside hoop only.)
- Acrylic paints in bright colors like he ones featured on the project shown
- Metal paint (if you want to change the colors of the clock hands)
- Water-based matte polyurethane
- Picture wire
- Eye screw
- 2 large D swivel ring hangers
- Clock movement and extra-large hands

DESIGNER: SHELLEY LOWELL

TOOLS

- Sandpaper
- Wood filler
- Epoxy glue and wood glue
- Jigsaw with a wood blade for fine cutting
- Artist's brushes in a variety of sizes
- Pencil
- Electric drill with various drill bits

INSTRUCTIONS

1 If you're using an ironing board top for your clock tower, you can leave it attached to its stand at first, making it a handy work surface. Screw the two pieces of wood measuring 1½ x 1½ x 8 inches (3.8 x 3.8 x 20 cm) onto the back of your clock tower. Position one across the middle (where you'll later attach D swivel rings for hanging the clock), and one about 5 inches (12.5 cm) up from the bottom.

2 Fill and sand any blemishes or cracks in the wood, and smooth all of the edges.

3 Glue the wooden blocks, balls, and numbers to the clock tower in a random arrangement. Be sure to leave spaces for the Slinkys, which you'll attach later.

4 Transfer the cat patterns on page 123 to the Masonite, and cut out one of each. Sand the edges of the cats, then glue them in place with wood glue.

5 Coat your clock tower and all of its embellishments with one or two coats of gesso.

6 Coat the inside embroidery hoop with gesso, then paint it a color of your choice.

7 Draw or trace a circle for the clock face near the top of the tower, and drill a hole in the center of the circle.

8 Let loose and paint. Work on the clock tower base first, dividing it in half with a freehand squiggly line and painting one side one color and the other side another, with the clock face a third color in the middle. On the rest of the pieces, alternate colors and create whimsical patterns of dots, dashes, swirls, and checks. You can use the patterns and colors in the design shown as a guide, or simply follow your instincts.

9 Once the paint is dry, let it cure for two days.

10 Remove the stand from the ironing board top, leaving the two horizontal wood pieces you screwed to the back in step 1 in place.

11 Coat all of the painted pieces with polyurethane, and let them dry.

12 Place the Slinky on the right side where you want it. Mark each end's position lightly with a pencil. Make tick marks on opposite sides of the outside of the top ring and the inside of the top ring, then repeat the marking process for the bottom ring. You will have two pairs of tick marks for each end of the Slinky.

13 Drill a tiny hole at each tick mark. (The holes need be only a bit larger than the picture wire you are using.) Cut four lengths of picture wire, each approximately 6 inches (15 cm) long.

14 Decide how you want the Slinky (the clock's sprung springs) to hang and how many rings you want held in place with the picture wire. Then, starting with one of the four pairs of holes, insert the picture wire from the back of the clock tower through one hole in the pair, bring it over one end of the Slinky ring(s), then thread the wire back through the second hole in the pair, and tie or twist the ends tightly. Repeat this process, working through the three pairs of holes.

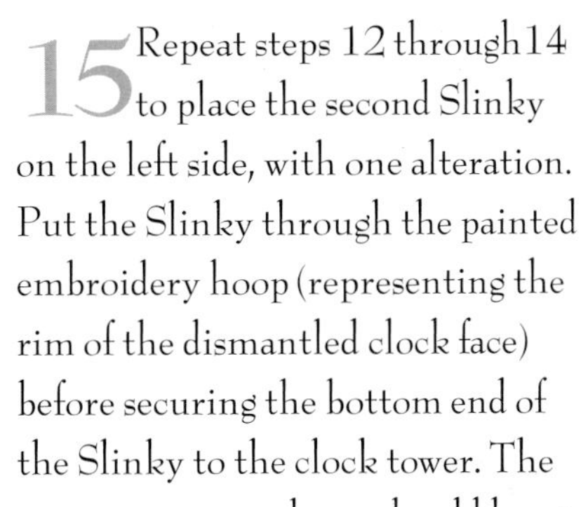

15 Repeat steps 12 through 14 to place the second Slinky on the left side, with one alteration. Put the Slinky through the painted embroidery hoop (representing the rim of the dismantled clock face) before securing the bottom end of the Slinky to the clock tower. The hoop should hang freely (or perhaps catch on a nearby block, like the one in the design shown). If you want one of the Slinkys to come around the side of the clock tower slightly, screw an eye screw into the back of the tower on that side, near the edge. Thread a short length of picture wire through the eye screw's hoop and around the ring that you want positioned at the side, then tie or twist the wire together, leaving enough slack so the Slinky hangs freely.

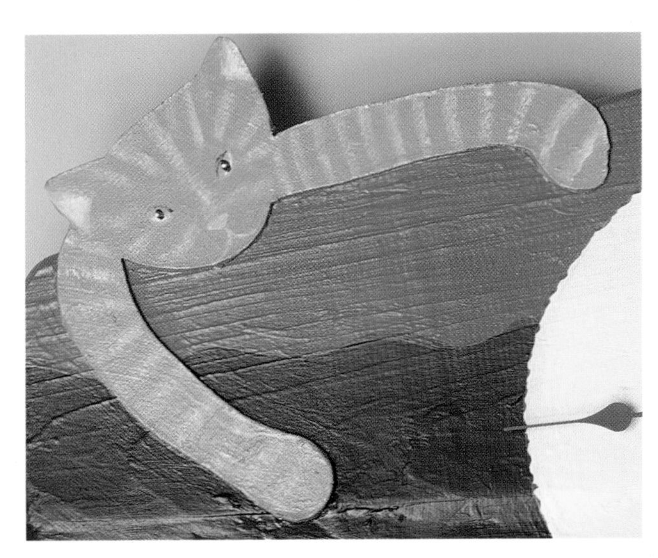

16 Paint the clock hands with metal paint, if you like, and let them dry.

17 Insert the clock movement and attach the hands.

18 Attach the two swivel ring hangers to the piece of wood you screwed to the middle of the back of the clock tower in step 1.

Subscribing to the Theory that You Can Never Have Too Much Time

Clock collector Perry Robinson has cuckoo clocks from the Black Forest in Germany, 100-year-old time clocks from American factories, clocks from World War II fighter planes, and clocks from Russian submarines. He's got clocks made with all-wooden gears held together with wooden pegs, clocks displayed under glass domes, clocks that are pieced-together puzzles, and a clock from the early 1800s with a pendulum suspended on a silk thread. They come from Germany, France, Switzerland, and other countries around the world.

His oldest clock dates to the early 1700s. The newest receives its signal from the atomic world clock in

Boulder, Colorado. There are more than 250 in all, covering the walls, shelves, and most surfaces of his Hendersonville, North Carolina, home. No one who lives there, including the dog, stirs when the orchestra of chimes, bells, gongs, and bird calls sounds every 15 minutes, every 30 minutes, every hour.

It all started 12 years ago, when Robinson inherited an antique clock and taught himself how to repair it. Now, in addition to his own ever-growing collection of working clocks, he's got a basement shop full of gears, springs, screws, and tiny tools—and heirloom clocks from around the region that he's bringing back to life for their owners.

A sense of history drives the obsession. He says he loves the challenge of making something 200 years old work again. And his personal collection of clocks—including those that have been through wars, those that told time in European schoolhouses, those that graced parlors in estate homes, and those that were lovingly trans-

ported across oceans and continents hundreds of years ago—grew out of his interest in the story behind each one.

"Sometimes, I just sit and wonder whose mantels these clocks used to sit on, and I think: 'If only they could talk'."

DESIGNERS: ALLISON AND TRACY PAGE STILWELL

Distressed Mantel Clock

CUT OUT THIS CLASSIC OLD CLOCK SHAPE. Then use a simple finishing technique to make it look like it's a well-worn treasure you snagged at an antique auction. In the end, you get a great clock for your mantelpiece without having to outbid anyone.

CUT LIST

DESC	QTY	MATERIAL	DIMENSIONS
Clock face	1	¾-inch (1.9 cm) pine	1 x 9½ x 23½ inches (2.5 x 23.8 x 58.8 cm)
Bases	2	¾-inch (1.9 cm) pine	1 x 2 x 5 inches) (2.5 x 5 x 12.5 cm)
Feet*	4	wooden ball knobs	1½ inches (3.8 cm) in diameter

*Wooden ball knobs with one flat edge are available at most craft stores. The flat edge is necessary for gluing the feet onto the bottom of the clock bases.

MATERIALS

- White acrylic primer
- Acrylic paints in burnt umber and off-white or pale yellow
- Candle
- 4 screws, 2 inches (5 cm) long
- Press-type numbers, in an antique font (Press type is available at stores that sell craft, art, and office supplies. It takes the stress out of printing and painting neat, consistent letters and numbers. Be sure to add several coats of sealer to protect press-type symbols.)
- Acrylic sealer
- Antique medium
- Clock movement and hands

TOOLS

- Pencil
- Ruler or measuring tape
- Protective eyewear
- Jigsaw
- Router with a piloted bit (for cutting decorative edges) of your choice
- Artist's brush, medium size
- Sandpaper, medium- and rough-grade
- Rasp (optional) (a hand-held woodworking tool used to contour edges and corners)
- Hammer
- Tack and/or nail, for hole punching
- Wood glue
- Drill, with bit to fit screws

INSTRUCTIONS

1 With the pencil, draw the outline of your clock shape on the large piece of pine, using the project photo as a guide. The clock featured rises from approximately 2¾ inches (6.9 cm) on each side to 9½ inches (23.8 cm) in the center.

2 Use the jigsaw to cut out the shape of the clock. We recommend you wear protective eyewear for this step and for step 3.

3 Rout the top and side edges of the clock shape and all sides of the base pieces.

4 Paint all surfaces of the clock pieces with two coats of the white acrylic primer. Sand the pieces as needed between coats. Allow the pieces to dry completely.

5 Paint some areas of the clock face, especially around the edges, burnt umber. Paint parts of the bases and feet (where you want them to look worn) burnt umber, as well. Let the paint dry completely.

6 Once the burnt umber is dry, rub the candle over the brown spots, leaving a thin coat of wax. You'll go back and sand these areas later to uncover the color, so make a mental note of where they are.

7 Paint all surfaces of the clock pieces with two coats of off-white or pale yellow. Allow the paint to dry between coats.

8 Once the top coat is dry, sand the entire surface of the clock with the medium-grade sandpaper. Give special attention to the areas that you covered with burnt umber paint and wax—the paint should come up easily from these areas. Don't be afraid to sand vigorously. Use rougher sandpaper or a rasp on the edges of the pieces.

9 To add to your clock's aged appearance, use the hammer to make marks and dings in the wood. Add the kind of character only "bug holes" provide by piercing the wood with a sharp tack and/or nails.

10 Once you're happy with the results of your sanding and banging, center the clock face on the bases and use wood glue to attach the pieces. Apply wood glue to each surface and let it set up slightly before attaching the pieces. Reinforce the bases with the 2-inch (5 cm) screws, drilling two screws up through the bottom of each base.

11 Apply wood glue to the flat surfaces on each of the feet, let it set up briefly, then attach two feet to each base piece. Allow the glue to dry completely before handling the clock.

12 Use the ruler and pencil to find and mark the center of the clock face. Apply the press-type numbers to the clock face, positioning them evenly around the center point. With the tack or nail, take nicks out of the press type to add to the clock's antique appearance. Afterwards, use a brush to apply several coats of acrylic varnish or sealer.

13 Following the manufacturer's instructions, apply the antique medium to all surfaces of the clock. Be sure to rub the medium into all the marks and holes.

14 Insert the clock movement and attach the hands.

A Little Pageantry with Your Time?

Surviving six centuries, the ornate astronomical clock mounted on the Old Town Hall in Prague, Czech Republic, continues to measure time and the movements of celestial bodies—and to delight observers with an elaborate hourly display.

Historians judge that the Orlaj, as the Czech clock is called, was created sometime around 1410, and is the result of a collaboration between a clock maker and a professor of mathematics and astronomy. While other European cities such as Strasbourg boast older astronomical clocks, the Orlaj is the only one still operating with its original clockworks. In its early years of use, the clock was highly regarded as a scientific observatory and source for astronomical data. Its original (and still-working) astronomical dial displays the movements of the moon, sun, and stars in relation to the earth. Observers can use the positions to determine the time of day or the day of the year and to predict equinoxes, eclipses, and phases of the moon.

Over the centuries, artists embellished the clock with an unusual collection of expressive statues, moving figures, and painted symbols. A skeleton represents death, a turbaned Turkish man signifies the invasions of earlier centuries, a man with a mirror warns against vanity, and a miser with a bag of gold embodies greed. They all contribute to the real draw of the Orlaj: an hourly pageant, in which the figures spring into action.

Crowds gather to watch the skeleton of death turn its hourglass upside down and toll a bell, then nod its head in the direction of the Turk, who looks away from death's gaze. The vain man looks at himself in the mirror while the miser shakes his bag of gold. Above the display, statues of Jesus Christ and the 12 apostles move before two small windows. Below, the archangel Michael passes judgement on the scene. At the end of the mechanical morality play, a rooster crows, and all the figures resume still positions.

Not surprisingly, the Orlaj also inspired a legend. It tells that after the clock was created, the clock maker, named Sir Hanus in the story, was sought after by officials from other cities, who wanted clocks just as fine for their towns. In response, the jealous powers in Prague ordered the village hangman to blind Sir Hanus, to prevent him from crafting another clock that might compete with theirs. Sir Hanus wanted revenge. He felt his way up the clock face and destroyed the clockworks, preventing the marvelous and beloved creation from keeping time for a hundred years.

Though there is no evidence the story is true, the Nazi assault on the city during World War II did take its toll on the clock. Thanks to the efforts of city residents, the clock components were repaired, and the entire creation has been restored to its original working form.

Painted Clocks

Wired

So, you thought pendulum clocks came in one style: staid. Here's a project that brings out the spirited side of a traditional base design.

Designers: Allison and Tracy Page Stilwell

MATERIALS

- Purchased clock base (For this project, the designers chose a plain wooden base, which includes the face and a stand with a slot for the pendulum. This base style is widely available in craft stores.)
- Acrylic craft paint in yellow, orange, pink, lavender, green, yellow-green, mint, blue, black, and white
- Watercolor paper (optional)
- Watercolor paints in the same colors listed above for acrylic paints (optional)
- 9 pieces of medium-gauge copper wire, each approximately 10 inches (25 cm) long
- Variety of beads
- Pendulum clock movement and hands

TOOLS

- Sandpaper
- Scissors or a craft knife
- Hole punch (optional)
- Permanent marker
- Variety of artist's brushes, such as wide flat, medium flat, and narrow round
- Peel-and-stick paper
- Drill and small bit for holes for wire
- Heavy craft glue

INSTRUCTIONS

1 Sand the clock and apply a base coat of white acrylic paint. When it's dry, sand the piece again.

2 Paint the clock with acrylic paints, using the color combinations shown or others you prefer. Using thin coats of slightly contrasting colors over one another will create more interesting final colors. To texturize the look, you can wipe off some areas of the top coat while the paint is still wet (making a patch of the color underneath visible), or lightly sand some areas after the paint is dry.

Here are the colors to use to replicate the project shown.
Face piece: orange and yellow on the outer circle, yellow in the center, blue and lavender around the edge, mint on the back.
Top of stand: yellow over orange on the surface, orange in the groove, white around the edges.
Legs: lavender.
Bottom of stand: blue over lavender on the surface, orange over yellow in the groove, white around the edges.
Feet: two yellow and two pink.

3 Add painted accents, such as thick lavender stripes around the edge of the face piece, thin black stripes around the edges of the stand, and vines and roses on the legs and surface of the bottom of the stand. Paint the vines and the leaves by alternating green and yellow-green. Paint the roses freehand or transfer small copies of the pattern on page 123 to use as guides. Use a mixture of pink and white paint, with yellow in the center.

4 To create the clock face, adjust the size of the sunburst pattern on page 123 on a photocopy machine until it fits the face piece of your clock (with a band of yellow visible behind the rays). Transfer the sized pattern to the watercolor paper, and cut it out. Paint the sun's rays in orange, yellow, pink, and lavender. Transfer the pattern for the grouping of roses on page 123 to the center of the face. Paint them as you did the roses on the stand, then fill in the background of the face with a light wash of blue on white. When the paint is dry, hand letter numbers on the face with the permanent marker. (You could also paint the face directly onto the wood. The watercolor method simply allows you to experiment with your design before finalizing it.)

5 Punch or cut a hole in the center of the completed clock face, and attach it to the wooden face piece with peel-and-stick paper.

6 Use the handle of an artist's brush to wrap the nine pieces of wire into tight spirals, leaving approximately 1½ inches (3.8 cm) at the end of each piece for adding beads and inserting the wire into the clock face.

7 Drill nine holes along the top edge of the clock, using a drill bit that matches the thickness of your wire. Center the first hole at the very top of the clock. Position the others out from there, about 1 inch (2.5 cm) apart.

8 Add several beads to the end of each piece of wire. Dip the end of each piece of beaded wire into the glue, and insert one in each hole. You may want to hold the wires in place as they dry.

9 Insert the clock movement and attach the pendulum and the hands.

Tribal Prints

EVERYONE WHO SEES IT (and asks what it's made from) is sure to have a why-didn't-I-think-of-that reaction. A hatbox lid turned clock—how perfect. Add some adventurous stamping and sponging, and it becomes a clock on the wild side.

DESIGNER: LYNN B. KRUCKE

MATERIALS

- Hatbox lid (The one used in the project shown measures 10½ inches [26.3 cm] long and 8 inches [20 cm] across.)
- Acrylic craft paint in white, black, yellow ochre, olive, rust, and bronze
- Paper towels
- Permanent black ink pad
- Clear acrylic spray
- Several inches (cm) of wire, approximately 26 gauge
- African-style beads
- Clock movement and hands

TOOLS

- Ruler
- Pencil
- Craft knife
- Sponge brushes
- Paint palette or plastic lid
- Masking tape

- ➢ Flat sponges (Available at craft stores, these sponges expand when they get wet.)
- ➢ Scissors
- ➢ Index card or piece of heavy card stock
- ➢ Spiral stamp
- ➢ Pushpin
- ➢ Craft glue

INSTRUCTIONS

1 Find and mark the center of the hatbox lid. Using the pointed blade of the craft knife, make a hole at this point that is large enough to accommodate the clock movement shaft. (Don't make the hole any larger than necessary; you want the shaft to fit snugly.)

2 Cover the lid (including the sides) with one to two coats of white acrylic paint, and let it dry.

3 Tear off a piece of masking tape long enough to go across the lid and sides diagonally. Press it to your clothes or to a dishtowel several times to remove some of the stickiness, then place it across the lid, a little off center, to create two sections, with the section on the top slightly bigger than the one on the bottom. Press the tape in place, making sure the edges are sealed well.

4 Paint the smaller section black, and let it dry. Leave the tape in place.

5 Tear off several more small strips of masking tape, then tear each small strip along its sides to create ragged edges that come to a point. You'll use these ragged strips to paint your zebra stripes. Remove some of the stickiness, then place them in a pattern you like on top of the black paint, and press them firmly to seal the edges. Use a clean, dry sponge brush to add a layer of white paint to the unmasked areas. As soon as possible, remove all the tape, pulling it up gently to avoid removing paint with it. Let the piece dry completely.

6 Paint the top section of the lid yellow ochre, and let it dry.

7 Cut the flat sponge into a couple of different shapes. (The designer of the project shown chose a triangle and a thick, curvy line). After cutting the shapes, wet the sponges so they'll expand, then squeeze out as much excess moisture as possible.

8 Add small amounts of olive, rust, and bronze paint to your palette or lid, fold the paper towels to create a blotter, and begin stamping the sponge shapes onto the yellow area. You can follow the pattern and color combinations shown or choose your own. First, dip one of the sponge shapes into the paint, being careful not to soak up too much. Blot the sponge on the paper towels, then press the shape onto the yellow surface. Don't worry about perfect images; the sponges will create a primitive, hand-stamped appearance that suits this project perfectly.

9 Dip the edge of the index card into one of the paint colors, then use it to add some fine, wavy lines to the yellow section.

10 Finally, stamp spirals onto the yellow section using black permanent ink.

11 Once the clock has dried completely, spray it with two or three coats of clear acrylic spray.

12 Mark the points on the clock face for 12, 3, 6, and 9 o'clock, and use the pushpin to make holes at each point. Thread a small section of wire through each bead. Attach a bead to each point by pulling the wire ends through to the back of the lid and twisting them to secure the bead. You may want to add a dab of glue to the back of the beads for additional strength and to keep them in position.

13 Attach the clock movement to the back of the lid with glue. Once the glue is dry, attach the hands.

I'M LATE, I'M LATE!

ALICE'S WHITE RABBIT—equipped with a hook on his hand—makes the world's most clever display piece for that heirloom pocket watch you've been keeping in a drawer. The pattern and paint tips provided here are all you need to create your own Wonderland watch stand.

DESIGNER: JEAN TOMASO MOORE

MATERIALS

- Piece of ¾-inch (1.9 cm) pine or plywood approximately 2 x 3 feet (.6 x .9 m) long
- Large sheet of transfer paper (optional)
- Piece of wood 8 x 12 inches (20 x 30 cm) for base
  2 strips of wood approximately ¾ x 12 inches (1.9 x 30 cm)
- 4 screws
- Wood putty
- 40 inches (102.5 cm) of decorative molding (optional)
- Finishing nails
- White pigmented water-based primer/sealer
- White latex enamel satin finish
- Acrylic craft paints in black, yellow, green, light blue, and pink
- Small eye hook
- Small picture hanger
- Fast-drying clear gloss polyurethane
- Pocket watch with chain

TOOLS

- Jigsaw
- Protective eye wear
- Pencil
- Screwdriver
- Drill
- Hammer
- Sandpaper
- Artist's brushes in various sizes
- Ruler
- Pliers
- Disposable foam brushes

INSTRUCTIONS

1 On a photocopy machine, enlarge the rabbit pattern on page 124 to the size you want. (You may have to enlarge it in sections, then tape the sections together.) The rabbit in the project shown stands about 30 inches (75 cm) high.

2 Transfer the pattern to the plywood either by cutting it out, laying it on the plywood, and tracing around it, or by placing a sheet of transfer paper between the pattern and the plywood, and tracing over the pattern.

3 Using the jigsaw and wearing safety glasses, cut the rabbit pattern out of the plywood.

4 Stand the rabbit cutout on the piece of wood you're using for your base. With the pencil, mark lines on the base to show where the front and back of the rabbit piece will stand.

5 Screw a ¾- x 12-inch (1.9 x 30 cm) strip of wood on the outside of each of the pencil lines. Predrill and countersink the holes to keep the wood from splitting. After the strips are screwed in place, fill the holes with wood putty. The rabbit cutout should slide between the strips and stand upright on the base (see figure 1).

6 If you like, you can add decorative molding to the edges of the base, using finishing nails to secure it.

7 Sand all the edges and rough spots on the wood, prime all the surfaces with white primer/sealer, and allow it to dry.

8 Apply a top coat of white latex enamel to the wood.

9 Once the top coat is dry, fill in the details of the rabbit design with a pencil, using figure 2 as a guide. If you like, you can enlarge figure 2 on a photocopy machine and use transfer paper to transfer the design to your rabbit

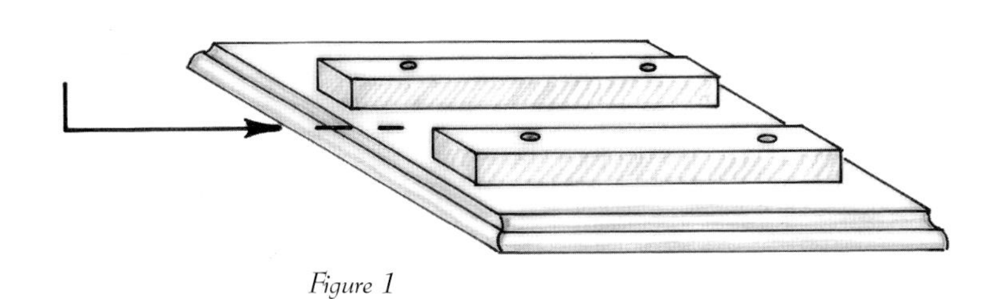

Figure 1

piece. Using a thin brush, outline the entire design with black paint. Once the black outline is dry, paint the jacket green, the vest yellow, and the buttons and the trim on the collar light blue. Use pink to highlight the ears. (You can substitute colors that will complement the room the rabbit will occupy, if you like.)

10 Paint a checkerboard design around the rabbit's feet and on the base. You can either paint the design freehand or mark the checks with a pencil and ruler first.

11 Determine where the pocket watch will look most natural in the rabbit's hand, and insert a small eye hook there. (Open the hook with pliers before screwing it in).

12 Attach the watch to the hook, and allow the chain to fall naturally. On the back of the rabbit, nail a small picture hanger to hold the tail of the watch chain out of sight.

13 When all your construction is complete, slide the rabbit out of the base and use foam brushes to apply a coat or two of high-gloss polyurethane to the front of all surfaces, then reassemble the construction.

Figure 2

Why (How, When, and Where) Does the Cuckoo Bird Sing?

Just why 18th-century German clock maker Franz Ketterer chose to equip his clock with the call of the cuckoo rather than, say, that of the hoot owl we may never know. But how, when, and where his first official cuckoo clock came to be are clock-enthusiast questions with answers.

Ketterer, working in the Black Forest, a range of low mountains in southwestern Germany, had quite a bit of clock-making company. The first clock built in the region was created by a winter-bound farmer in his work shed in 1640. Made entirely of wood, it was known as a "wood-beam" clock. Soon, other snowed-in farmers took to making clocks during the long winter months, finding clock sales an ideal way to supplement their incomes. Eventually, Black Forest clock makers spent the summer months trekking across Europe, clocks strapped to their backs, selling their finely carved crafts. Word of the region's talented artisans quickly spread; by the mid-18th century, clock making had grown into a thriving industry in the Black Forest.

This house-size structure has been dubbed by those in its Black Forest hometown of Schonach as the "Earth's Largest Cuckoo Clock"—surely they're right!

In 1730, Ketterer outfitted one of his clocks with an ingenious innovation. He replaced the customary chime that signalled the half hour and hour on most Black Forest clocks with two different bellows that worked together to imitate the natural call of the cuckoo bird. And the cuckoo clock was born.

Black Forest cuckoo clocks still feature the charm, rich detail, and fine hand carving that sparked their original popularity—with a few modern technological twists. Many now play musical selections in addition to sounding the celebrated bird call. The traditional German songs "Edelweiss" and "Der Frohliche Wanderer" are frequent picks, as are the popular Oktoberfest titles "Trink, Bruderlein, Trink" and "In Munchen steht ein Hofbrauhaus." Mechanical moving parts, from dancing couples and chopping woodcutters to turning waterwheels, also grace the front of many modern-day cuckoo clocks.

Flying Furniture Clock

MOST DAYS, it seems the old adage about time taking wing and soaring along is true—whether you're having fun or not. Here's a great way to transform a standard piece of furniture into a reminder that you might as well enjoy the flight.

DESIGNERS: ALLISON AND TRACY PAGE STILWELL

MATERIALS

- 2 pieces of plywood, each measuring approximately ½ x 5 x 9 inches (1.3 x 12.5 x 22.5 cm) for angel wings (If you want your wings much larger than those in the project shown, you'll need larger pieces of plywood.)
- Compact disk cabinet (The one in the project shown is 6½ x 8½ x 46½ inches [16.3 cm x 21.3 cm x 1.2 m].)
- White acrylic primer
- ¼-inch (6 mm) plywood, 3½ x 8½ inches (8.8 x 21.3 cm), for clock face
- Acrylic paints in an assortment of colors (To replicate the color scheme in the project shown, use light gray, purple, light lavender, light yellow, bright yellow, bright orange, light green, yellow-green, blue-green, light blue, light turquoise, black, and medium blue.)
- Playful-style press type for numbers (Press type is available at stores that sell craft, art, and office supplies. It takes the stress out of printing and painting neat, consistent letters and numbers. Be sure to add several coats of sealer to protect press type symbols.)
- Acrylic sealer
- Decorative drawer pull or cupboard handle, approximately 1¼ inches (3 cm) in diameter x 2 inches (5 cm) deep
- Screw, sized to fit drawer pull/cupboard handle and long enough to go through the clock face and the front facade of the roof peak
- Clock movement and hands

TOOLS

- Jigsaw
- Artist's brushes, fine tipped, medium tipped, ¼-, ½-, and 1-inch (6 mm, 1.3, and 2.5 cm) flat brushes
- Medium-grade sandpaper
- Ruler
- Pencil
- Wood glue
- 2 books or other heavy objects (optional)
- Drill with ¼-inch (6 mm) drill bit

INSTRUCTIONS

1 Enlarge the pattern for angel wings on page 124 to a size that suits your design. (The wings on the project shown are ½ x 4 x 8 inches [1.3 x 10 x 20 cm] each.)

2 Transfer the pattern to the two pieces of plywood designated for the angel wings, and use the jigsaw to cut out the wings.

3 With an artist's brush, apply one coat of white acrylic primer to all of the surfaces of the compact disk cabinet, the clock face, and the wings. Let the primer dry completely. Use the sandpaper to sand the parts as necessary.

4 Apply two coats of paint to the cabinet and wings, using an assortment of colors to create a fun look that will complement your home decor. Allow the paint to dry completely between each coat, then sand the pieces as necessary.

To recreate the color scheme featured here, paint the cabinet as follows (mixing colors to create various shades, as necessary): roof, light mint green; shelves, light gray; inner cabinet sides, light green-blue; inner cabinet back, medium blue; outside of cabinet, light yellow; roof peak front facade, purple; roof edge, light green-blue.

5 Once the cabinet is dry, use the ¼-inch (6 mm) flat brush to decorate the front of the roof edge with a checkerboard design, applying a freehand line of lavender paint about every ¼ inch (6 mm). Also, add a checkerboard design to the front edges of the cabinet, this time using the ½-inch (1.3 cm) flat brush to apply a freehand line of yellow-green every ½ inch (1.3 cm).

6 Use the fine-tipped brush to write "Time Flies" across the

Got a Minute?

It's a question people often ask before proceeding to take up a half an hour of your time. But if you really do have just 60 seconds to spare, here's a starter list of 25 productive (or at least spirit-lifting) ways you could spend them.

1. Brush your teeth
2. Pay a bill
3. Drink a glass of water
4. Look through a roll of 24 pictures
5. Change a light bulb
6. Give a friend a hug
7. Make a paper airplane
8. Throw a Frisbee several times
9. Replace a roll of toilet paper
10. Write a postcard
11. Soften a ball of clay
12. Lose at a game of pinball
13. Pick up several pieces of trash
14. Flip through a catalog
15. Share a kiss
16. Make someone smile (see number 15)
17. Warm a cup of coffee in a microwave
18. Water a plant
19. Eat a pickle
20. Make a first impression
21. Draw a picture
22. Look up a word in a dictionary
23. Cut your nails
24. Thread a needle and take several stitches
25. Do a cartwheel

front facade of the roof peak, if you like. The design shown features purple letters thinly shadowed with black.

7 Paint the fronts of the wings bright yellow, the backs of the wings medium blue, and the edges of the wings yellow-green. When the surfaces are dry, use the fine-tipped brush to apply bright orange dots to the front of the wings and medium blue dots to the edges. Let the wings dry completely.

8 To attach the wings to the cabinet, lay the cabinet facedown on a stable work surface. On the back of the cabinet, measure 1½ inches (3.8 cm) down from the top of one side and make a mark. Repeat this measurement on the other side. The tops of the wings will be placed at these marks. Apply an ample coat of wood glue to the front of each wing base (where the wing will be attached) and to the back of the cabinet (where the wings will be attached). Allow the wood glue to set up for several minutes, then press the wings in place and allow them to dry completely. You may want to set a book or other heavy object on top of each wing to help secure the bond.

9 Paint the plywood clock face with several coats of light turquoise, allowing it to dry completely between coats. Paint a decorative border around the edge of the face in light lavender, using a flat brush and simple strokes to create the triangular shapes shown in this design. Allow the clock face to dry completely.

10 Use the ruler to find the center of the clock face and mark the spot. Drill a ¼-inch (6 mm) hole for the clock movement at this center mark.

11 With press type, apply numbers on the clock face, being sure to position them evenly around the center point. With an artist's brush, apply two coats of acrylic sealer to protect the numbers.

12 Insert the clock movement and add the hands.

13 Position the clock face at the top of the cabinet, allowing the roof peak front facade to overlap the face. Experiment with it until you find an angle you're happy with, then drill a hole in the top of the clock face. Use the drawer pull or cupboard handle and a screw to attach the clock face to the front facade of the roof peak.

Talisman of Time

THIS DELIGHTFUL LITTLE sprite is a good-luck charm and timepiece in one—offering you not only the hour, but the sun, the moon, and the stars.

DESIGNERS: ALLISON AND TRACY PAGE STILWELL

MATERIALS

- Purchased wooden clock base, 4½ x 9 inches (11.3 x 22.5 cm)
- Acrylic craft paint in turquoise, blue, yellow, lime, and white
- Polystyrene foam ball (The one in the project shown is approximately 2 inches [5 cm] in diameter.)
- Aluminum foil
- Paper-based modeling clay
- 13½ inches (33.8 cm) of ¼-inch (6 mm) dowel
- 3 strips of fabric about 2½ inches (6.25 cm) wide and at least 20 inches (50 cm) long
- Metallic thread
- Small amount of yarn for hair (You can substitute other materials such as commercial doll hair or raffia.)
- Several inches (cm) of high-gauge wire
- Bezel clock movement to fit the hole in the base

TOOLS

- Drill and ¼-inch (6 mm) drill bit
- Woodcarving tools
- Sandpaper
- Medium or large flat brush
- Glue gun and glue sticks or craft glue
- Sculpting tools
- Craft knife
- Needle or pin
- Hair dryer (optional)
- Hack saw or old wire cutters

INSTRUCTIONS

Body

1 Drill holes approximately ½-inch (1.3 cm) deep on the top edge of the clock base at 10, 12, and 2 o'clock.

2 Carve a simple design into the wooden base. The project shown features a patchwork-quilt design, with the squares filled with stripes, squiggles, and checkerboards. Two grooves run up the middle of the base, coming to a point at the top; the area inside the grooves is filled with dots. When you're finished carving the design, sand the clock to remove any splinters.

3 Give the whole piece a base coat of white paint, making sure you get paint into all the grooves.

4 Paint the different sections of your carved design in a variety of colors, keeping the paint out of most of the grooves. When the paint is dry, sand it to smooth the surface, removing some of the paint for a mottled look. Go over some of the sections with a slightly different color to give the piece a bit more interest. Lightly sand the surface again. You can repeat this process as many times as you like. (Remember, it's just paint. If you don't like the combination, you can wipe the section you're working on with a damp rag, sand it, or paint over it.)

Head and Accessories

1 Securely wrap the polystyrene foam ball with foil. You may want to make a "handle" where the corners come together at the back.

2 You'll create the doll face out of two layers of clay. First, roll a 2-inch (5 cm) ball around in your hands, then press it into a flat piece about ¼ inch (6 mm) thick. This piece will cover the front half of the foam ball; don't make it any larger. If it covers more than half the ball, you won't be able to remove it in step 9.

3 Next, make a second layer of two "slabs," each about ⅛ inch (3 mm) thick, one for the forehead and the other (which should be slightly larger) for the area from the lower eyelids to the chin. (See figure 1.) Smooth the two slabs onto the first layer. Water will help smooth the transition points. Leave a small space between the two top slabs to make the eye sockets.

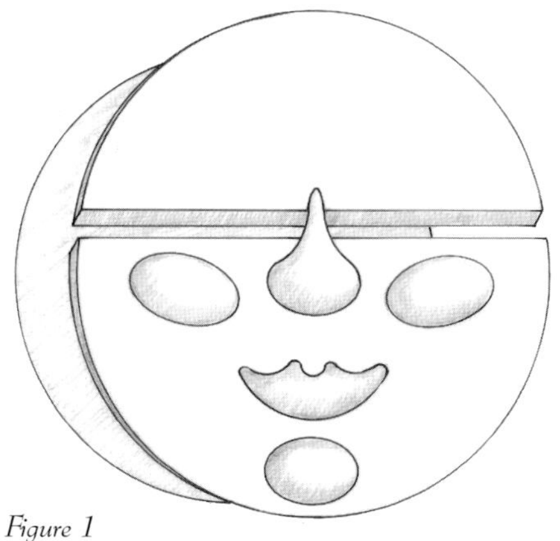

Figure 1

4 Form a small triangle for the nose, using your thumbs and forefingers together. Place the top portion of the nose over the eye line and smooth it into the surface. Use a small sculpting tool to make nostrils.

5 Apply two small circles of clay, flattened, under the eye area to make cheeks. Smooth and shape them.

6 Roll two very small "snakes" for the lips. Put them in place and smooth the edges to connect them to the face. If you're not pleased with them, simply pull them off and try again, until you've created an expression you like.

7 Place a small ball of clay under the mouth to build up a chin. Add or take away as much clay as necessary.

8 You can manipulate the eye area and add little balls for eyeballs, if you like.

9 Make sure all the transition areas are smooth and seamless. Carefully remove the face (all layers together) from the foam ball, and let it air dry; a sunny windowsill can help the drying process. You want the clay to be completely dry (this should take a couple of days) before you start to paint.

10 Prepare some small, flat circles of clay and cut out several stars, a moon, and a sun with the craft knife. Use a needle or pin to make a hole in each piece, then let the clay dry.

11 Once the pieces are dry, you can use fine sandpaper to smooth their surfaces.

12 Use several thin layers of blue and white paint to create a celestial scene (complete with clouds and stars) on the dried face. A hair dryer is helpful for drying the paint between coats. (Paper-based clay is subject to damage from mold, mildew, and bugs. With that in mind, make sure you add a coat of paint to the inside of the head as well as the face to help protect it.) Paint the stars, sun, and moon yellow.

Assembly

1 With the hack saw or old wire cutters, cut two pieces of dowel rod measuring 5½ inches (13.8 cm) each and one piece measuring 2½ inches (6.25 cm). Wrap each of the longer pieces of dowel (which will become the arms) with a strip of cloth, starting with the "hand" end. Leave ½ inch (1.3 cm) at the other end of each for attaching the arms to the clock.

2 Make a hole where the neck should be by inserting the shorter piece of dowel into the base of the head about 1 inch (2.5 cm). Remove the dowel, fill the hole partially with a sturdy craft glue, then insert the dowel again and let the glue dry.

3 Wrap the head with a portion of the remaining fabric square, and stitch around the base of the ball to secure the fabric (stitching it to itself), leaving enough loose fabric to later wrap the neck. Add glue to the drilled neck hole on the top of the clock, insert the other end of the neck dowel, and let the glue dry.

4 Once the head is secure, wrap the remaining fabric around the neck. Secure the fabric with glue or with a needle and thread.

5 Add glue to the arm holes on the clock, and insert the wrapped arms.

6 Glue the painted face to the front of the head.

7 Sew small bundles of yarn onto the head.

8 Wrap the wrists with wire bracelets, and stitch the stars, sun, and moon to the hands with metallic thread.

9 Insert the bezel movement.

Folk Art Clock

SELECT SOME SIMPLE images. Paint them in an unschooled style. When you finish, you've got a homespun clock with irresistible, back-to-basics charm.

DESIGNERS: ALLISON AND TRACY PAGE STILWELL

MATERIALS

- Purchased wooden clock base similar in shape to the one in the project shown (The project clock measures 3/4 x 10 x 15 inches [1.9 x 25 x 37.5 cm] and came with a predrilled hole for the clock movement.)
- Acrylic paints in assorted colors (such as light jade green, dark green, olive green, gold, red, creamy white, sky blue, black, raw sienna, and dusty pink)
- Glaze paint, medium blue
- Clock movement and hands

TOOLS

- Artist's brushes, fine tipped, small, and medium
- Medium-grade sandpaper
- Comb, for painting
- Pencil

INSTRUCTIONS

1 Using the medium-sized brush, paint all surfaces of the wooden clock base creamy white. Let the base dry completely, then sand it as needed.

2 Paint the outside edge of the clock base dark green and the routed trim around the edge gold. Let the paint dry completely.

3 Paint the front edge of the clock base light jade green, and allow it to dry completely. Once the front edge has dried, carefully apply a thick coat of the blue glaze over the light jade green. Use the small comb to comb out the glaze, starting at the inside of the edge and moving outward. Let the glaze dry for at least four hours.

4 Paint the groove around the clock face and lower panel red. (If your base has no groove, simply paint a ⅛-inch-wide [3 mm] border around the clock face and the lower panel.) Apply several coats of creamy white paint to the clock face, allowing it to dry completely between coats.

5 To create the farm scene featured on the lower panel, use the pencil to lightly sketch a few hills on the horizon, about halfway down the panel. Then, use simple shapes to draw the house and barn and to add a few trees and small animals. Paint the farm scene in assorted colors of your choice, then allow the panel to dry completely.

6 With the small and fine-tipped brushes and an assortment of acrylic paints, decorate the clock face with characters from your farm scene or other folk-art elements (quilt squares, pitchers, simple flowers, etc.) that appeal to you, placing one image at each number space. You may want to paint actual numbers for the 3, 6, 9, and 12 o'clock locations. Allow the clock to dry completely.

7 Insert the clock movement and attach the hands.

Art Girl

SURE OF HERSELF. Unafraid to make an artistic statement. Lovably quirky and unpredictable. Sounds like a dream of a dinner-party guest. Not to mention the world's most wonderful best friend. Why not have her hanging around all the time?

DESIGNERS: ALLISON AND TRACY PAGE STILWELL

MATERIALS

- Purchased clock base, large gothic triptych
- Wood scraps* for body parts, in the following dimensions:
  - 2 legs, one 12 inches (30 cm) long and one 16 inches (40 cm) long
  - 2 arms, both 10 inches (25 cm) long
  - 4 or 5 head pieces, each 5 to 6 inches (12.5 to 15 cm) long
  - 2 feet, each 5 to 6 inches (10 to 12.5 cm) long
- Acrylic paints in assorted colors (such as purple, lime green, black, white, antique rose, orchid, yellow-green, periwinkle blue, crocus yellow, and gold)
- 2 small hinges, ¾ inch (1.9 cm), packaged with screws
- 4 wooden hearts, approximately 1½ inches (3.8 cm)
- Rubber stamp and ink pad (optional)
- Clock movement and hands

TOOLS

- Sharp craft knife or other cutting tool, for cutting wood scraps
- Artist's brush
- Medium-grade sandpaper
- 1-inch (2.5 cm) foam paintbrushes
- Craft glue
- Screwdriver

*The designers used scraps of old plaster lathe scavenged from the attic to create their Art Girl body parts. If you don't have a handy scrap stash, check local lumberyards for leftovers. Prepare the body parts using a sharp craft knife, a razor blade, or other sharp cutting tools. Don't be fussy about how evenly they're cut—unevenness and other imperfections add to the clock's charm. Select a number of odd pieces with pointy ends to use for the head. To make the feet, cut two triangular points off the ends of a scrap of wood.

INSTRUCTIONS

1 Apply one or two coats of white acrylic paint to the triptych. Let it dry completely, then sand the surface as needed. Paint the back of the triptych periwinkle blue, the front of the center purple, and the front of the sides lime green.

2 With the 1-inch (2.5 cm) foam brush, apply stripes of white paint about every inch (2.5 cm) to the legs and arms. When the white paint has dried completely, paint 1-inch-wide (2.5 cm) black stripes between the white stripes.

3 Paint the feet antique rose, and decorate them with dots of purple. Let them dry completely.

4 Paint each of the head pieces a different color—these designers chose yellow-green, orchid, purple, and yellow. Once the paint is completely dry, glue the pieces together in a pile so they resemble a star. Be sure to leave one piece long enough to serve as a neck. Let the head dry overnight.

5 Glue the head (by the neck piece) and the arms to the triptych, and glue the feet to the ends of the legs. Attach the hinges to the tops of the legs and then to the triptych, positioning them about 1½ inches (3.8 cm) in from each side.

6 Decorate the four hearts with tiny checkerboards of black and white paint around the edges. Paint the surface of each heart a different color, let them dry, then glue them around the clock face at the 3, 6, 9, and 12 o'clock positions.

7 Decorate the clock face with painted-on gold stars and the body with any added accents you like. These designers used a rubber stamp and ink pad to print question marks on the doors.

8 Apply several coats of white paint to the clock hands, letting them dry completely between coats. Decorate the hands with black dots and stripes. Insert the clock movement and attach the hands.

Message Center

FORGET THE SCATTERED system of covering the refrigerator with magnet-held reminder scraps while casting frantic glances at the clock on the other side of the kitchen. Here's a clever and efficient center that helps you keep it all together.

MATERIALS

- Medium-density fiberboard (MDF), smooth pine, or plywood, 10½ x 24 inches (26.3 x 60 cm)
- Blackboard paint (available at most paint stores)
- Acrylic paints in assorted colors, such as light green, yellow-green, blue-green, red, yellow, and creamy white
- Watercolor paper or heavy card stock, for clock face (This design features a 7-inch-square [17.5 cm] clock face.)
- Acrylic polyurethane sealer, gloss or satin
- 1 piece of "L"-shaped molding, 5½ inches (13.8 cm) long, for chalk ledge (optional)
- Length of ribbon, for hanging chalk (optional)
- Clock movement and hands

DESIGNER: KIMBERLY HODGES

TOOLS

- Sanding block
- Sponge roller
- Artist's brushes
- Ruler
- Pencil
- Scissors
- Craft glue
- Drill and bit to fit the shaft of your clock movement
- Hand saw
- Wood glue

INSTRUCTIONS

1 Lightly sand the sides and face of the large piece of wood with the sanding block.

2 Roll one coat of blackboard paint onto the front of the board with the dry sponge roller. After the first coat is completely dry, add a second, thicker coat. Press lightly with the sponge roller to avoid leaving roller tracks or paint ridges on the board. Light pressure will help create a smooth, even blackboard finish. Allow the blackboard paint to dry completely.

3 Paint a scalloped border along the outer edges of the board. (Actually, any decorative border will work if you prefer something else. Just be sure to leave room for the clock face in the center of the board and for messages underneath.)

4 Paint the back of the board the same color as the border. Allow the paint to dry completely.

5 Create your clock face design on heavy card stock or watercolor paper. Draw and paint your own design or use the project photo as a guide. You could also create a collage for your clock face, gluing snippets of magazine images, fabrics, and other materials onto the paper and/or stamping images with a rubber stamp, then sealing them with the acrylic sealer.

6 Use the ruler and pencil to measure and mark the center of the clock face. Also, lightly mark where you want to place the clock face in the upper center of the board.

7 Cut out the clock face design, and sponge roll a thin, even coating of craft glue onto the back of the paper. Don't panic if the paper begins to roll up, just work quickly! Using the palms and sides of your hands and working in rapid, circular motions, rub the clock face design down onto the blackboard. Begin in the center of the paper and work out to the sides. Continue to rub the design down for two or three minutes. This technique typically removes any air bubbles that might form under the paper.

8 Let the glue dry completely (at least 15 minutes), then use an artist's brush to apply a light coat of the acrylic polyurethane sealer to the clock face design. Allow the sealer to dry for at least 15 minutes, then apply a second coat.

9 Drill a hole in the blackboard through the center of the clock face.

10 Insert the clock movement and add the hands.

11 To add a chalk ledge to your blackboard clock, use the saw to cut a 5½-inch-long (13.8 cm) section from the piece of "L"-shaped molding (this piece can be longer if you like). Place the molding along the bottom edge of the clock body, positioning the upper edge of the molding behind the body. Use wood glue to attach this edge of the molding to the back of the clock body. Allow the glue to dry completely before handling the clock.

VARIATION

To make your blackboard clock chalk-friendly in a different way, drill a small hole in one lower corner of the blackboard. Thread a fun piece of ribbon through the hole and tie it in place, then tie a piece of chalk (in a complementary color, perhaps) to the other end.

Paper, Fabric & Clay

Remember When

Here's a great way to turn all those routine glances at the hour and minute hands into reminders of some of life's most important moments.

Designers: Allison and Tracy Page Stilwell

MATERIALS

- Black-and-white copies of favorite photos (If you weren't planning for this project when you snapped great color shots years ago, worry not. The black-and-white setting on a color copy machine produces a good-quality reproduction of originals, whether they're color or black and white.)
- Black matting, 12½ x 16 inches (31.3 x 40 cm), with a 5- x 7-inch (12.5 x 17.5 cm) opening
- Access to a computer (optional)
- Sheet of white paper
- Black frame 12½ x 16 inches (31.3 x 40 cm)
- Rub-on numbers
- Clock movement and hands

TOOLS

- Scissors or craft knife
- Glue stick
- Ruler
- Pencil
- Drill and ¼-inch (6 mm) drill bit
- Awl

INSTRUCTIONS

1 After assembling and copying your photos, cut them apart and arrange them in a collage fashion on the matting.

2 Use the glue stick to carefully attach them to the matting, working with the pieces on the bottom first.

3 Begin to create your clock face. You can print the phrase "Remember when . . ." onto the white paper from a computer. Be sure to position the phrase so it prints near the bottom of the page and isn't wider than the opening in your matting. If you don't have a computer, you can hand print the phrase, cut and paste letters from a magazine, or use rub-on lettering. Position the paper behind the matting so the phrase is centered along the bottom of the opening, glue it in place, and temporarily add the frame backing behind it.

4 The top 5 square inches (12.5 square cm) of the paper are where you'll position the numbers and hands of your clock face. Determine the center of the section (2½ inches [6.25 cm] down and 2½ inches [6.25 cm] in), and mark it lightly with the pencil. On the backing, carefully measure to find the corresponding spot and mark it. When you're sure you've coordinated the points, remove the backing and drill a hole where you made your mark. Use the awl to carefully poke the hole you marked in the paper (coming through from the back of the paper).

5 Apply the numbers to the clock face.

6 Insert the clock movement and attach the hands. Insert the clock into its frame.

Pressed Leaf Clock

WHAT A WONDERFUL paradox. First, you get to reach back to the simple, childhood pleasures of squeezing, shaping, and pressing things into clay. When you're finished, you end up with a sophisticated, grown-up clock.

DESIGNER: LYNN B. KRUCKE

MATERIALS

- Black polymer clay (This project will take about three blocks.)
- Leaves (Collect an assortment or pick just one type.)
- Gold pigment ink pad
- Water-based glaze made for using with polymer clay
- Clock movement and hands

TOOLS

- Wax paper
- Pasta machine (one you can devote to polymer projects), dowel, or rolling pin for rolling the clay
- Craft knife
- Ruler
- Drinking straw
- Pencil
- Baking tray
- Craft glue

INSTRUCTIONS

1 Working on wax paper, condition the clay by kneading it and rolling it in your hands to warm and soften it.

2 Use the pasta machine, dowel, or rolling pin to roll a slab of clay 3/16 inch (5 mm) thick. With the craft knife, cut a rectangle 3½ inches (8.8 cm) wide and 4¾ inches (12 cm) tall. Remove the excess clay.

3 Press the leaves gently onto the gold ink pad, face up, to coat the backs with ink. (Don't worry if you don't get total coverage, the charm of leaf printing is that no leaf is perfect, so no print need be perfect either.) Place the leaves on the clock face randomly, and press them carefully to create impressions and transfer the ink to the clay slab. The ink won't dry until the piece is baked, so be careful not to smear the ink when you remove the leaves.

4 Find the center along the top of the rectangle, then measure down 1¾ inches (4.5 cm) to mark the center of the clock. Use the drinking straw to poke a hole for the clock movement, wiggling the straw to make the hole large enough. Be careful to keep the hole round, and don't let the clay get thicker around this edge. Smooth the clay, if necessary, to keep a ridge from forming around the hole.

5 Use the eraser end of a pencil to make impressions at the 12, 3, 6, and 9 o'clock positions.

6 Carefully transfer the piece (with the wax paper still underneath it) to a baking tray, and bake it according to the directions on the clay package. Allow the piece to cool completely before moving on to the next step.

7 Once the piece is cool, apply one or two coats of a glaze suitable for use with polymer clay. (The wrong glaze will react with the clay and leave the piece sticky.)

8 Recondition the remaining clay, if necessary, and roll it into a thick log about 3 inches (7.5 cm) long. Press and smooth the piece into a rough triangular shape that will serve as a holding base for your clock, then flatten it somewhat. (The shape isn't as important as making sure the piece will support the clock face.)

9 Holding the clock face firmly, press it into the clay base near the front to create a groove where the clock will sit. Remove the face carefully and set it aside (if you leave it in place at this point, it will stretch out the shape of the base). Gently press the clay from the back of the base forward to build up the area right behind the groove, where the face will lean, but be careful not to narrow the groove.

10 Smooth away any fingerprints and bake the base, again following the manufacturer's directions. Glaze the base once it's cool.

11 Insert the clock movement and attach it to the back of the clock, using a strong craft glue. Let it dry overnight, then attach the hands.

Paper Moon

IF CREATING SOMETHING OUT OF LAYERS of decorative, handmade paper is your idea of heaven, here's a celestial-theme clock to make your heart soar. Who would have guessed that the simple process of cutting and pasting could lead to this?

DESIGNER: CLAUDIA LEE

MATERIALS

- Cardboard for templates
- Sheet of heavyweight card stock (80-pound [36 kg] cover-weight stock works well) that is slightly larger than the finished clock you have planned (A sheet of card stock measuring 8½ x 11 inches [21.3 x 27.5 cm] would be perfect for the project shown.)
- Assortment of decorative, handmade papers
- Acrylic gloss medium
- Tube of dimensional paint (fabric paint is a good choice) in a color that coordinates with your paper colors
- Small length of nylon fishing line
- Clock movement and hands

TOOLS

- Craft knife
- Pencil
- Cutting mat (optional) (You can also use a thick magazine.)
- Craft glue
- Newspaper or wax paper
- Artist's brush (1 inch [2.5 cm])
- Cotton swabs (optional)
- Straight pin

INSTRUCTIONS

1 Transfer the patterns on page 124 to the cardboard.

2 With the craft knife, cut out the cardboard shapes, which will serve as templates. Work on top of the cutting mat, a magazine, or something else to protect the surface underneath from the blade of your craft knife.

3 Use the templates to trace the moon and star shapes onto the heavier card stock. These pieces will become the backing for your clock.

4 Use the templates, again, to trace one of each shape onto your decorative paper.

5 Cut the shapes out with the craft knife, and glue the paper shapes to the backing.

6 Repeat the process of tracing and cutting out paper shapes and gluing them in place (especially important if you're working with very thin papers) until you're happy with the layers of colors and textures.

7 Lay your clock pieces on the newspaper or wax paper, and brush each with a coat of acrylic gloss medium. Allow the acrylic to dry completely (checking the pieces as they dry to make sure they aren't sticking to the newspaper or wax paper), then brush each with a second coat. Once the second coat is completely dry, apply a single coat to the back of each piece, and let the acrylic medium dry.

8 Use the craft knife to carve out a small circle where you want the center of your clock face to be.

9 With the dimensional paint, outline and decorate the clock pieces. In the project shown, the designer outlined each piece, added dashes around the perimeter of the moon, squeezed out numbers to create a dial around the hole cut in step 8, and added tiny pairs of dashes to the surface of the star. If you mistakenly add a dab of paint where you don't want it, simply dampen a cotton swab and wipe it off.

10 Once the paint is completely dry, poke a tiny hole in the tip of the moon and at the top of one of the star's points. Use a short piece of fishing line to connect the two pieces, knotting it off on each end so that the star hangs nearly even with the face of the clock.

11 Insert the clock movement and attach the hands.

Clay Rainbow

IF YOU'VE BEEN LOOKING FOR AN EXCUSE to roll up your sleeves and join the polymer clay craze, here's a perfect first project. Start with a purchased wooden clock base in this standard, arched shape, then cover it with a patchwork of brightly colored triangles. Couldn't be easier, and the result is pure playfulness.

DESIGNERS: ALLISON AND TRACY PAGE STILWELL

MATERIALS

- Polymer clay in green, red, blue, black, and white
- Semicircle wooden clock base
- Bezel clock movement to fit the hole in the base

TOOLS

- Wax paper
- Pasta maker (one you can devote to polymer projects), dowel, or rolling pin for rolling the clay
- Craft knife
- Baking sheet

INSTRUCTIONS

1 Use a piece of wax paper to cover your work surface. Begin kneading the basic clay colors together in various combinations to come up with about a dozen shades, each approximately the size of a walnut.

2 With the pasta machine equipped with its thinnest roller (or using your dowel rod or rolling pin), roll out the clay to about 1/8-inch (3 mm) slabs (the thickness of medium cardboard).

3 With the craft knife, cut out freehand triangles measuring about 1 inch (2.5 cm) on each side.

4 Lay the clock on its back and, beginning at the bottom, place a row of clay triangles onto the clock base, overlapping them slightly. Gently press the pieces together.

5 Continue pressing rows of triangles into place until the front is covered. You may want to work with various shades of a single color on each row, as the designers did on the project shown. Trim any excess clay at the edges of the base.

6 With the leftover clay, cut strips measuring 1½ inches (3.8 cm) long and ¼ inch (6 mm) wide. Apply the strips along the arched top of the wooden base. Again, you will want to overlap them slightly so they stick together, and you may want to work in gradations of shades.

7 Make two long "snakes," one white and one with black, beginning each with a 2-inch (5 cm) ball of clay. Hold them together, then twist and stretch them together until they resemble a barber-shop pole. Roll them on a flat surface to help integrate, smooth, and lengthen them. You need to end up with a black-and-white strip that is 36 inches (91.5 cm) long.

8 Gently but firmly press the black and white trim onto all of the edges of the clock base.

9 Create a small black-and-white snake and place it around the center hole, giving yourself enough space to insert the bezel movement. Use the craft knife to trim the snake, if necessary.

10 Lay the clock on its back on a piece of wax paper on a cookie sheet, and bake it according to the instructions on your package of polymer clay. Pay attention to timing; the clay can burn and darken if it's left in the oven too long. Let the clock cool.

11 Insert the bezel movement into the center hole.

Brown Bagging It

SAVE WHAT YOU would have spent eating a sandwich out today. Take the money and buy yourself a clock movement. You've got the bag you packed your lunch in to use as your base. Only things left to add are the design on the front and a bit of weight in the bottom. Efficient, inexpensive, and easy!

DESIGNER: BARBARA BUSSOLARI

MATERIALS

- Two small sheets of card stock in coordinating colors
- Brown paper lunch bag
- Black marker (and other colors, if you like)
- Construction paper in assorted colors
- Fabric (optional, to embellish clock face character)
- Number stamps
- Mat board, approximately 3½ x 7 inches (8.8 x 17.5 cm)
- Buttons and/or beads in various shapes, sizes, and colors
- Cat litter, 1 cup (140 g)
- Clock movement and hands

TOOLS

- Ruler
- Pencil
- Scissors
- Craft glue
- Craft knife
- Ink pad
- Cutting board or a magazine
- Awl
- Hot glue gun and glue stick (optional)

INSTRUCTIONS

1 Use the ruler to measure the dimensions of one side of the paper bag. Measure and cut rectangles from two colors of the card stock to fit the front of the bag. One should be slightly larger, so it provides a border for the other. (On the project shown, the large rectangle measures 4½ x 7 inches [11.3 x 17.5 cm], and the smaller one measures 3½ x 6 inches [8.8 x 15 cm].)

2 Glue the smaller rectangle to the larger rectangle, creating an even border between each shape. Let the glue dry completely.

3 With the black marker, draw a line inside the border of the larger rectangle, ⅛ inch (3 mm) away from the outer edge of the smaller rectangle. To add a wavy effect to the line, use the ruler as a guide but shake the marker slightly as you go.

4 Using a pencil and working on the thinner construction paper, draw your design for the clock centerpiece. (You can use the patterns on page 124 as guides or come up with your own design.) Cut out the centerpiece design with the scissors or craft knife, and glue it in place on the card stock. Add any details with the black (or any other color) marker.

5 Use the number stamps and ink pad to add numbers to the clock face. Remember to space the numbers evenly if you plan to position numbers around the entire face, and make sure the numbers 6 and 12, and the numbers 3 and 9, sit directly across from one another on the clock face.

6 Locate and mark the center point on the back of the clock face. Position the card facedown on the cutting board or magazine, and use the awl to punch a hole through the center mark. With the hole as a guide, use the craft knife to carefully cut a circle large enough to accommodate the clock movement. To smooth the edges of the hole, push a round pencil or marker through the opening from the front of the card.

7 Measure and cut a piece of the mat board to a size approximately 2 inches (5 cm) smaller than the clock face. The mat board will fit inside the paper bag to help support your clock movement. Find the center of the mat board, and

repeat step 6 to punch a hole for the clock movement.

8 Position the clock face on the front of the paper bag. Trace the hole in the center of the clock face onto the bag, then cut it out with the craft knife.

9 Hold all of the pieces in place (the mat board inside the bag and the face on the front), then insert the clock movement and attach the hands.

10 Add beads, buttons, or other baubles to the clock face, gluing them in place with craft glue or a hot glue gun. Make sure to position them so that they do not interfere with the movement of the clock hands.

11 Pour approximately 1 cup (140 g) of cat litter into the paper bag to help it stand up. After installing the batteries in the clock movement, fold the top of the bag down as you would a lunch bag.

VARIATIONS

- Incorporate fabric as clothing for the centerpiece of the clock face.
- Use a scanner or color photocopier to create an image from a photograph (your child, a friend, etc.) for the clock face.

Snapshot Clocks

IF A PICTURE IS WORTH a thousand words, there's no telling how much time a photo will buy you. What's certain is the irresistible appeal of transforming your own into big, bright clocks.

DESIGNERS: ALLISON AND TRACY PAGE STILWELL

MATERIALS

- Photograph, enlarged to a size you like
- Foam core*
- Peel-and-stick mounting adhesive (available at craft-supply stores)*
- Clock movement and hands

TOOLS

- Sharp craft knife

*Note: The amount of foam core and peel-and-stick mounting adhesive you need will depend on the size of your picture. These clocks were made from photographs blown up to 8½ x 11 inches (21.3 x 27.5 cm).

INSTRUCTIONS

1 With the craft knife, cut the foam core to fit the size of your enlarged photograph.

2 Carefully apply the peel-and-stick mounting adhesive to the front of the foam core.

3 Use the craft knife to cut your photo to its approximate final shape (cutting around the edges of specific images, etc.).

4 Peel off the second side of the mounting adhesive and place your photograph on top of it. Position the photograph carefully; it will be difficult to adjust it once it's made contact with the adhesive.

5 Use the craft knife to cut out the final shape, trimming closely around the edges of the photograph and foam core.

6 Determine where you want your clock hands on the photograph, mark the spot, and use the craft knife to cut a hole for the movement shaft.

7 Insert the clock movement and attach the hands.

That's One Big Clock

IF YOU WERE THROWING A PARTY for a crowd of urban loft-dwellers who work in architecture, photography, and interior design, you could hang this clock and make them all feel right at home. Of course, the simple, spare design (not to mention the size) are appealing whatever your social circles.

DESIGNERS: ALLISON AND TRACY PAGE STILWELL

MATERIALS

- Foam core board, 36 inches square (90 sq cm)
- Clock face
- Peel-and-stick mounting adhesive (Available at craft stores. The benefit of this material is that it's moveable until you burnish it.)
- Black acrylic paint (optional)
- High-torque clock movement (If you can't find them at your local craft supplier, high-torque clock movements are available through clock-part catalogues.
- Large clock hands (Those on the clock featured measure 22½ inches [56.3 cm] and 18 inches [45 cm]. Chances are, you'll need to special order hands this big through your local craft store or from one of the suppliers listed on page 128.)

TOOLS

- Access to a photocopy shop
- Glue stick
- Ruler
- Pencil
- Craft knife with sharp blade
- Artist's brush (optional)

INSTRUCTIONS

1. Take the clock face to a full-service photocopy shop and have it enlarged to approximately 34 inches (85 cm) in diameter.

2. Carefully apply the mounting adhesive to the back of the foam core, moving it until it's positioned where you want it.

3. Remove the backing from the adhesive mounting, recruit the help of another pair of hands, and carefully position the enlarged clock face on the sticky surface. Working out from the center, press the face onto the board, one side at a time.

4. After rubbing out any creases in the clock face, follow the adhesive manufacturer's instructions to make your placement permanent. Use the glue stick to attach any edges that don't adhere completely.

5. Use the ruler and pencil to lightly mark the center of the clock face, then use the craft knife to cut a hole at the center point.

6. With a sharp craft knife blade, cut around the clock face, trimming off the corners of the foam core board to form a large circle. Leave a border of approximately ¾ inch (1.9 cm) around the clock face.

7. Insert the clock movement and attach the hands.

VARIATION

Leave a border of approximately 1 to 1½ inches (2.5 to 3.8 cm) around the clock face, then use the artist's brush and the black acrylic paint to add a painted edge.

Three-Panel Picture Frame Clock

THESE DAYS, SEEMS EVERY ONE OF US is trying to accomplish several things at once. Here's a tripod clock design that picks up on the trend. Picture frames on two sides. Beaded clock face on the third. What end table wouldn't feel productive with this three-panel piece on its surface?

DESIGNER: KIM TIBBALS-THOMPSON

MATERIALS

- Several pieces of heavy cardboard
- 1/3 yard (.3 m) each of two coordinating fabrics (This designer used tapestry and linen.)
- 12 decorative beads to mark the hours
- 2 2/3 yards (2.4 m) waxed cotton cording or other cording (embroidery floss could be used) cut in 4-inch (10 cm) lengths
- 1 1/3 yards (1.2 m) of suede lacing or other cording
- 4 beads to slip on the ends of the suede ties
- 2 decorative buttons on shanks
- Clock movement and hands (Make sure the movement shaft is long enough to go through two thicknesses of cardboard and three thicknesses of fabric, for a total of approximately 1 1/4 inches [3 cm].)

TOOLS

- Craft knife
- Ruler
- Sewing scissors
- Iron
- Pencil
- Needle and thread
- Awl or punch tool
- Heavy duty spray adhesive
- Masking tape
- Glue gun and glue sticks
- Straight pins

INSTRUCTIONS

1 Cut the cardboard into the following pieces: three panels measuring 6 x 9 inches (15 x 22.5 cm); two mats (for picture frames) measuring 4-1/2 x 6 inches (11.3 x 15 cm); one clock face measuring 4 1/2 inches (11.3 cm) square.

2 Cut a 3- x 4-inch (7.5 x 10 cm) opening in each of the 4 1/2- x 6-inch (11.3 x 15 cm) mats, leaving a 3/4-inch (1.9 cm) margin on the top and sides and a 1 1/4 inch (3.2 cm) margin on the bottom.

3 Cut the tapestry fabric (which will cover the panels) into three pieces, each measuring 9 x 12 inches (22.5 x 30 cm).

4 Cut the linen fabric as follows: two pieces measuring 6 1/2 x 8 inches (16.3 x 20 cm) to cover the mats; one piece measuring 7 1/2 inches (18.8 cm) square to cover the clock face; three pieces measuring 7 x 10 inches (17.5 x 25 cm) to line the back sides of the three panels. With the iron, press each of the panel-liner pieces under 1/2 inch (1.3 cm) on all four sides.

5 Use the pencil to mark the linen clock face. First, locate the center of the fabric, where the clock shaft will be inserted, and mark it. Next, mark the locations of the hours.

6 With the needle and thread, sew the 12 beads in place to mark the hours on the linen clock face.

7 Use the awl or punch tool to punch a hole in the center of the cardboard clock face, where the clock shaft will be inserted. Make sure the hole is large enough for the clock shaft to enter easily.

8 Apply a heavy coat of spray adhesive to the back of the linen you cut for the clock face and, aligning the center holes, mount the fabric on the cardboard clock face. To secure the 1 1/2 inches (3.8 cm) of fabric that extend beyond the cardboard clock face, wrap the four corners onto the back of the cardboard, then wrap the sides around and press them down. Wrap the fabric tightly, but don't stretch it excessively. Hold the linen in place until the spray adhesive bonds.

9 Wrap each corner of the face with two parallel strips of the waxed cotton, and secure the cut ends with masking tape on the back side of the clock face.

10 To cover the mats, apply a heavy coat of spray adhesive to the back side of the mat fabric pieces. Lay the fabric facedown and center the cardboard mat pieces on

the fabric. Wrap the fabric around the cardboard, following the procedure described for the clock face in step 8. Allow the spray adhesive to bond before proceeding.

11 Continue working from the back of the mat pieces to form the center openings on the mats. Use the craft knife to cut the fabric on one mat from the center of the mat to each of the four corners (the cuts will form a big "X" in the center of the fabric). Pull each of the four points, which should still be wet with spray adhesive, to the back of the mat and hold them in place until the spray adhesive bonds. Trim away the excess fabric. The adhesive will prevent the corners from fraying. Repeat this process for the second mat. Wrap each corner with the waxed cotton cording, as described in step 9.

12 To make the three panels, apply a heavy coat of spray adhesive to the back side of the pieces of panel fabric. Lay the fabric facedown and center the cardboard panels on the fabric. Wrap the fabric around to the back of the cardboard, using the procedure described for the clock face in step 8.

13 Assemble the clock. Place the three panels side by side vertically, fabric side up, on your work surface. On the center panel, mount the clock face with the glue gun, positioning it ¾ inch (1.9 cm) from the top and each side. Using the awl or another punch tool, pierce a hole in the panel where the clock shaft will protrude.

14 On the left panel, position a mat ¾ inch (1.9 cm) from the bottom and sides, with the widest margin of the mat at the bottom. Cut two 12-inch-long (30 cm) pieces of suede lacing and tuck two of the cut ends (about ½ inch [1.3 cm] worth) between the mat and the panel on the lower left side of the mat, about ¾ inch (1.9 cm) up from the bottom edge of the mat (you'll use these laces to help tie the panels together later). Hot glue the mat to the panel along the top, left, and bottom edges of the mat, gluing the suede lace ends in place, as well. Leave the right side of the mat open so you can slide in a photograph.

15 On the right panel, position the remaining mat, widest margin at the bottom, so that it is ¾ inch (1.9 cm) from the top and sides of the panel. Cut two 12-inch-long (30 cm) pieces of suede lacing and tuck two of the cut ends (about ½ inch [1.3 cm] worth) between the mat and panel on the upper right side of the mat, about ¾ inch (1.9 cm) down from the top of the mat edge. Hot glue the mat to the panel along the top, right, and bottom edges of the mat, gluing the suede lace ends in place, as well. Leave the left side open so you can slide in a photograph.

16 Slip beads on the ends of the four pieces of suede lacing, and knot them in place.

17 To attach the lining and the lace "hinges," turn the center clock panel over. Turn the left panel over and align it on the right side of the center panel. Turn the right panel over and align it on the left side of the center panel. Place the panels about ⅛ inch (3 mm) apart. Cut the remaining suede lacing into eight 3-inch-long (7.5 cm) pieces. Position two pieces of lacing across the gap between the left and center panels, 1½ inches (3.8 cm) below the top edge of the panels. Position two pieces of lacing across the same gap, 1½ inches (3.8 cm) above the bottom edge of the panels. Tack the raw ends in place with hot glue. Repeat the procedure on the gap between the center and right panels. Center the lining fabric over the back sides of each panel, covering all the raw edges of the lacing and fabric, and hot glue it in place. Using the awl or punching tool, pierce a hole through the center panel lining to accommodate the clock shaft.

18 Sew a button on the front side of the left panel, centered over the mat, about 1½ inches (3.8 cm) from the top edge of the panel. Sew a button on the front side of the right panel, centered beneath the mat, about 1½ inches (3.8 cm) from the bottom edge of the panel.

19 Insert the clock movement in the center panel and attach the hands.

20 Stand the panels up on their bottom edges, fold the left and right panels back to form a three-sided structure, and secure the panels by lashing suede lacing with beaded ends (the pieces you created in steps 14 through 16) around the buttons.

Clock of the Rising Sun

This late-for-work sun seems to be hitting the snooze button with one hand and half-heartedly pulling himself up over the horizon with the other. If you took one look at this clock and felt as if you were staring groggily into a mirror, this is the polymer-clay project for you.

DESIGNER: IRENE SEMANCHUK DEAN

MATERIALS

- Gessoed Masonite (available at art- and craft-supply stores), at least 6 x 8 inches (15 x 20 cm)
- 4 ounces (112 grams) of deep blue or purple polymer clay, for sky background
- Texturing materials (crumpled wax paper, lace, textured leather, sandpaper, etc.)
- Wax paper or plastic wrap
- 1 ounce (28 grams) of bright yellow polymer clay, for sun face
- Assorted polymer clay canes (including a black-and-white checkerboard polymer cane) and other embellishments (See the box on page 87 for information on making your own polymer canes. You can also find premade canes in most craft-supply stores.)
- Polymer clay in various colors (dark blue, for hand; translucent, for stars; and others that strike your fancy)
- Glitter
- 4 ounces (112 grams) of light blue translucent polymer clay, for columns
- Polymer clay varnish
- Clock movement and hands

TOOLS

- Jigsaw or scroll saw
- Craft knife
- Ruler
- Pencil
- Drill with drill bit to fit your movement shaft
- Polyvinyl acetate (PVA) white glue (This glue is compatible with polymer clay, and it's also heat resistant and anti-yellowing.)
- Pasta maker (one you can devote to polymer projects), dowel, or rolling pin for rolling the clay
- Wet sponge
- Tiny star cutter
- Tiny cutters (a drinking straw will work)
- Manufactured polymer clay hand mold
- Baking tray
- Cyanoacrylate glue
- Toothpick
- Small artist's brush

INSTRUCTIONS

1 Using the jigsaw or scroll saw, cut the gessoed Masonite into a house shape measuring approximately 5¾ inches (14.4 cm) wide and 8 inches (20 cm) high from the base to the roof peak.

2 Find and mark the center point of the house shape, and drill a hole large enough to accommodate the shaft of the clock movement.

3 With your fingertips, spread a light, even coat of the PVA glue over the gessoed side of the Masonite. Let it dry completely. This will provide a polymer-clay-friendly work surface on which to mount your clock (polymer clay will not stick to the gessoed Masonite otherwise).

4 Roll out a sheet of conditioned deep blue or purple polymer clay to an approximate ⅛-inch (3 mm) thickness, using the pasta machine or a hand rolling tool. Lay the clay sheet gently onto the backing, pressing outward from the center to get rid of any air bubbles. Choose from among your texturing materials to add some texture to the surface of the clay. In this design, crumpled wax paper was straightened out and pressed into the polymer clay. To keep the polymer from sticking to the texturing object, dampen the object slightly with a wet sponge before applying it to the clay surface.

5 Embellish the clock with polymer clay elements. As you apply each element, press it gently but firmly into the backing, using a piece of plastic wrap or wax paper over your finger to prevent fingerprints.

To recreate this design, hand form the bright yellow polymer clay into a half circle with a sleepy face, and position it at the bottom edge of the clock face. Using the rolling pin, roll out nine "snakes" of bright yellow polymer clay, and position them so that they radiate away from the face in wiggly shapes. Cut slices of assorted millefiori canes (human and animal faces, leaves, flowers, or other creations) and place them in various spots around the piece. Remember that you can either purchase pre-made canes or create your own following the techniques described on page 87. Using the tiny cutter, cut star shapes out of translucent polymer clay mixed with glitter (or silver polymer clay). Using a manufactured hand mold and dark blue polymer clay, create a polymer clay hand. Apply the hand to the design near the roof peak, so it appears to be sprinkling a handful of the silver stars into the sky.

6 Poking through with the pencil from the drilled hole in the back of the clock, make a small hole in the clay background where the clock movement shaft will protrude. Use the drinking straw to cut out a round hole around this spot on the front. Cut out tiny pieces of polymer clay (in shapes such as circles and squares) in a color that will stand out from the background, and position the pieces on the center hole to represent numbers.

7 Check your piece for stray lint and fuzz, which are often attracted to polymer clay, and remove any stowaways. Then, place the clock on the baking tray and bake it in the oven for 20 minutes at the clay manufacturer's recommended temperature.

8 With your hands, roll a snake of polymer clay to a ¾-inch (1.9 cm) thickness. Cover the snake with slices of the checkerboard cane. Roll the cane-covered snake on your work surface until the seams disappear. The cane-covered snake will need to be approximately 6 inches (15 cm) long for the bottom edge of the clock and 8 inches (20 cm) long for the roof edge. Dab several drops of cyanoacrylate glue along the Masonite edge at the bottom of the clock base, and spread it with a toothpick. Then, place a length of the checkerboard-covered snake along the edge and press it gently but firmly into place. Trim off any excess clay from the ends with the craft knife. Repeat this procedure along the roof edge at the top of the Masonite piece to apply the remaining checkerboard strip. Hold each strip in place for about 30 seconds while the cyanoacrylate glue dries.

9 To complete your clock, add polymer columns to the sides of the house shape. Begin with a light blue translucent polymer clay. (This designer mixed in glitter for a more phosphorescent effect.) Form two strips approximately 7½ inches (18.8 cm) long. Roll each strip to a ¾-inch-thick (1.9 cm) diameter. Apply a layer of cyanoacrylate glue to the edge of each column and place one column along each side of the clock, allowing about 2 inches (5 cm) to protrude above the top edge of the house

sides. Hold each column in place for about 30 seconds while the glue dries. Curl the tops of the columns outward slightly by hand. Add a few cane slices of flowers or other items to embellish the columns, if you like.

10 Bake the piece in the oven for 30 minutes at the clay manufacturer's suggested temperature. Turn off the oven, prop the door open, and allow the clock to cool for at least 30 minutes.

11 Once the clock is completely cool, you may want to apply a coat of varnish to the number markers to make them more visible. To make the clock hands stand out, use a small artist's brush to apply a light coat of varnish, then sprinkle a generous helping of glitter on each hand. Apply a second coat of varnish to the hands after the first is dry.

12 Insert the clock movement and attach the hands.

CANING WITH POLYMER CLAY

Caning, considered one of the most popular methods for working with polymer clay, stems from a Venetian glassmaking process known as "millefiori," or "one thousand flowers." Using this technique, a glassmaker bundles a selection of colored glass rods together to create a design. Similarly, a polymer clay artist compresses an assortment of colored clay sheets and shapes together to create a design. To visualize the structure of a cane, imagine a jelly roll: viewed from the side, it looks like a long yellow cake; viewed from a sliced end, it looks like a red-and-yellow spiral of jam and cake. Polymer canes follow the same premise: they may look like humdrum logs of clay from the side, but reveal wondrous designs ranging from delicate flowers to funny faces on their sliced ends. One of the best parts about the process is that, once you make an assortment of canes, you can keep them on hand and simply cut off slices for use in your designs.

Here's how to make a cane in a simple checkerboard design.

1 Choose two colors of conditioned clay, and use a roller (or pasta machine) to flatten each color into a rectangular sheet about 1/8 inch (3 mm) thick.

2 Stack the two sheets (if one is larger than the other, place the smaller sheet on top), and use a cutting tool to trim the edges of the stacked sheets to make one long rectangle.

3 Cut the rectangle into quarters, and stack the pieces to form a block of stripes with alternating colors.

4 Cut the striped rectangle into strips, then flip every other one upside down to form a checkerboard pattern.

5 With your fingers, compress (but don't stretch) the cane from one end to the other to work out any air bubbles and to convert the many layers into one piece of clay. This is called "reducing the cane."

6 Gently tug on one end of the compressed block while simultaneously rolling lightly (with a roller, if you're making a square or rectangular cane, or with your fingers, if you're making a round cane) on the other end. Work up and down the length of clay to elongate it evenly and to keep the ends from becoming too distorted. Trim off the distortion on the ends until you see a clear design.

You can use canes to form faces, flowers, and many other elements, as well. But if you don't feel up to making your own canes just now, you'll be delighted to know that a variety of pre-made polymer canes are available in most art- and craft-supply stores that carry polymer clay.

Face of Faux Ivory

YOU (ALONG WITH ANY NEARBY ELEPHANTS) will be relieved to know you don't have to drum up a single tusk to create this imitation-ivory clock. The creamy-white color and finely etched texture are the result of combining some clever techniques with polymer clay.

DESIGNERS: ALLISON AND TRACY PAGE STILWELL

MATERIALS

- 4-inch (10 cm) square of 1/4 inch (6 mm) plywood
- Acrylic craft paint in black, white, and raw sienna
- Small blocks of polymer clay in white, beige, and translucent paper
- Clock movement and hands

TOOLS

- Drill and small bit
- Fine sandpaper
- Medium flat artist's brush
- Small round artist's brush
- Container for water
- Rag
- Wax paper
- Pasta maker (one you can devote to polymer projects), dowel, or rolling pin for rolling the clay
- Craft knife
- Pointed sculpting tool, such as a toothpick
- Rubber stamps of numbers
- Baking sheet
- Glue

INSTRUCTIONS

1 Drill a hole for the clock movement in the center of the plywood.

2 Sand and prime the plywood base with white paint. Let the paint dry, then sand the piece again.

3 Cover the edges with additional coats of paint until they're well coated. (You don't need to worry about the front of the plywood piece, since it will be covered with clay.)

4 Using watered-down raw sienna, paint a thin coat over the white on the edges of the plywood. You're not trying to cover the white paint, only to create a glazed look. You may want to have a rag handy for wiping off excess paint.

5 Cover your surface with wax paper and begin kneading together the three colors of polymer clay: one part white (half a block), one part beige (half a block), and two parts translucent (whole block). You want to mix them fairly well, but it's fine to leave tiny bits of the original colors for interest. (If you have a pasta machine, you can just put chunks of each color through together, fold them on themselves, then put them through again until they're mixed.)

6 With the pasta maker, dowel, or rolling pin, roll the mixed clay into a 4-inch (10 cm) square about 1/8 inch (3 mm) thick. Trim the edges.

7 Using your pointed tool, draw a simple design in the clay. The project shown features a few flowers surrounded by a spiraling border and dots, for example. With the same tool, make small indentations along the edge of the clay square. Use rubber stamps or your pointed tool to add the numbers. (You can also use rubber stamps for adding designs to your piece.) Finally, poke a 1/4 inch (6 mm) hole in the middle of the square for the clock movement.

8 Transfer the square and wax paper to a baking sheet, and bake the clay according to the manufacturer's directions.

9 When the baked clay is completely cool, sand it lightly.

10 With water and a damp rag within reach, apply black paint to the baked clay and wipe it off immediately. You want the paint to soak into the sculpted grooves, but wipe off the surface. If necessary, use a combination of wiping and sanding to get the black paint into the places where you want it.

11 Apply thinned-down raw sienna to the surface of the clay in the same fashion, again, sanding areas (such as the corners) where you want the paint to adhere and create a worn look.

12 Glue the clock face to the plywood backing.

13 Insert the clock movement and attach the hands.

Dangling Ladies

THE WORD GETS SUCH A BAD RAP. We're not supposed to leave people, processes, or participles dangling. But the swinging legs on these playful ladies (who you can create from family photos, magazine images, paper dolls, or your own drawings) are what give them their kicky charm.

DESIGNERS: BRUCE ALLISON AND JILL MAYBERG

MATERIALS

- Sketch pad
- Inspirational materials such as magazines, photographs, or paper dolls
- Illustration board (at least 3-½ inches (8.8 cm) square to accommodate clock movement)
- Acrylic paints in a variety of colors
- Acrylic medium, gloss or matte finish (available at art-supply stores)
- ¼-inch-diameter (6 mm) jump rings (available at hobby, craft-supply, and jewelry-supply stores)
- Clock movement and hands

TOOLS

- Pencil
- Craft glue
- Artist's brushes (small detail brush and larger brush)
- Blow dryer (optional, to dry paints quickly)
- Sharp scissors
- Craft knife
- Hole puncher
- Pliers
- Hammer
- 5⁄16-inch (8 mm) metal hole punch for leather
- Cutting mat (or a couple of magazines)

INSTRUCTIONS

1 Draw your design in pencil on the watercolor paper or illustration board, or glue down a photocopied image or cut-out figure.

2 Paint the design with the acrylic paints, one color at a time. Use a small artist's brush for details and a larger brush to coat bigger areas of the design. Acrylic paints can be watered down or made more translucent with acrylic medium (a substance made up of the same plastic base as acrylic paints, but that lacks the colored pigments). You can mix the medium into your paints to achieve different effects; play with mixtures of various proportions to see what you like best. Allow the paint to dry completely after each coat (using a blow dryer set to no or low heat will do the job quickly, if you're feeling pressed for time). Then, paint the next color or embellishment on the design.

3 Once you're satisfied with your design and the paint is completely dry, use the large brush to apply several coats of the acrylic medium to the front and back of your clock figure. This will provide a strong plastic coating to protect the clock's surfaces and make the body more durable. Allow the medium to dry completely between each coat, and allow the final coat to dry for at least one hour.

4 Using the sharp scissors or the craft knife, cut out the clock's body, legs, and feet. Paint the edges of the illustration board black (or any other color you choose), and allow them to dry completely.

5 With the hole puncher, punch two holes at the bottom edge of the body and one hole in the top of each leg. Using the pliers, attach the legs to the body with the jump rings. Repeat this procedure to attach the feet, if you've cut them out as separate images.

6 Decide where you want the clock face to be, and mark the center point. Using a hammer and the 5⁄16-inch (8 mm) leather hole punch and working on the cutting board, punch a hole through the clock figure where you made the mark. (If you're fresh out of leather hole punches, you could also use a craft knife to carefully cut a hole for the clock shaft.)

7 Insert the clock movement and attach the hands.

Word of Encouragement: The nice thing about acrylic paints is that if you aren't happy with your work, you can simply paint over the old, dry paint for a fresh start. In fact, more layers usually makes for a more interesting finished piece, and top layers of paint can be scraped away to create a textured surface.

Common & Found Objects

Tin Tile Clock

Designers: Allison and Tracy Page Stilwell

THE STILWELLS SCORED THEIR TIN ceiling tile while walking through the neighborhood on trash day. If you're not quite so lucky, try scrap metal yards and companies that specialize in renovation. A bit of snipping and bending, a little faux rust, and this antique urban clock is yours.

MATERIALS

- Tin ceiling tile or similar metal square measuring approximately 10 inches (25 cm) square
- Instant iron and rust (Available at craft stores, these products work together to create a rusty effect.)
- Clock movement and hands

TOOLS

- Tinner's snips
- Work gloves
- Steel wool
- Protective glasses
- Drill and ¼-inch (6 mm) drill bit
- Small artist's brush

INSTRUCTIONS

1 Wearing work gloves, use the tinner's snips to make two 1-inch (2.5 cm) cuts at each corner of the tile. (See figure 1.)

2 Use a firm edge (such as the edge of a deck or of a protected table or counter) to bend down the 1-inch (2.5 cm) edges you created in step 1. Fold them onto the back of the tile.

3 With the steel wool, sand the edges of the tile. You might also want to use it to remove any flaky paint from the surface.

4 Put on the protective glasses, and drill a hole at the center of the tile.

5 Paint the clock hands with the instant iron, then the instant rust, following the manufacturer's directions.

6 When the hands are dry, insert the clock movement and attach the hands.

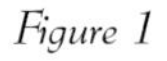

Figure 1

Rustic Wreath Clock

WOOD SMOKE. Flannel blankets. Rough-hewn furniture. This clock. Hang it on a rusty nail on the porch or mount it over the fireplace inside. Then sit back, watch it, and wonder what could be better than keeping time according to the twirling of twigs.

DESIGNER:
JEAN TOMASO MOORE

MATERIALS

- 8-inch (20 cm) diameter grapevine wreath (or any size of your choice)
- Foam core (large enough to fit inside diameter of wreath)
- Piece of handmade paper (preferably with leaves, flower blossoms, or other natural elements embedded in the surface)
- Double-sided foam tape
- 4 small leaves (silk or dried)
- Straight twig approximately ¼-inch (6 mm) in diameter and approximately 9 inches (22.5 cm) long
- 2 thin pieces of bark (for the hour and minute hands)
- Tiny forked twig (for the second hand)
- Small acorn (artificial or real)
- 1 yard (.9 m) sheer wire ribbon
- Artificial butterfly (optional)
- Picture-hanging wire, approximately 10 inches (25 cm)
- Clock movement and hands

TOOLS

- Scissors
- White craft glue and foam brush (for applying it)
- Ruler
- Awl
- Pencil
- Hot glue gun and glue stick
- Wire cutters

INSTRUCTIONS

1 Using the scissors, cut a piece of foam core to fit the inside diameter of the grapevine wreath.

2 Cut a piece of the handmade paper to cover the foam core.

3 Glue the handmade paper to the foam core with craft glue, creating the clock's face.

4 Use the ruler to find the center point of the clock face. With the awl, poke a hole large enough to accommodate the clock movement's shaft.

5 Use two strips of double-sided foam tape to adhere the clock movement to the back of the paper-covered foam core.

6 Arrange the four dried or silk leaves at the 3, 6, 9, and 12 o'clock positions on the clock face. Glue the leaves in place with the craft glue.

7 Cut and glue down small pieces of the straight twig to mark the remaining number spots on the clock face. You may want to lightly mark the locations in pencil before gluing the twigs in place.

8 Glue the thin pieces of bark on top of the existing hour and minute hands and the tiny forked twig onto the second hand.

9 Attach the hands, then glue the small acorn (or another natural ornament of your choice) to the center of the clock mechanism, on top of the screw that secures the second hand.

10 Attach the clock face to the wreath using the hot glue gun.

11 To embellish the wreath, tie a length of sheer wire ribbon in a large bow and tack it into place on the wreath with a dab of craft glue. If you like, glue on an artificial butterfly as well.

12 Use the wire cutters to cut a length of the picture-hanging wire. Thread the wire through the back of the wreath to create a secure wire loop for hanging the clock.

High-Tech Hardware Clock

STANDARD SHELF BRACKETS hold up each side, with a strip of metal screen in between. Aluminum flashing dotted with appliance surface guards forms the face. A quick trip to the hardware store, and you'll have just what you need to make this modern-look clock.

DESIGNER: JEAN TOMASO MOORE

MATERIALS

- 12-inch (30 cm) heavy-duty hanging shelf standard
- Piece of rolled aluminum flashing, approximately 6 x 14 inches (15 x 35 cm)
- 2 18½-inch (46.3 cm) black shelving brackets
- Approximately 2 feet (.6 m) of aluminum gutter guard (hinges removed)
- Approximately 22 inches (55 cm) of 18-gauge copper wire
- 2 wooden beads, approximately 1 inch (2.5 cm) in diameter
- Acrylic craft paint in turquoise, copper, black, and metallic ruby
- Spray acrylic sealer
- Piece of aluminum step flashing (used in roofing), 5 x 7 inches (12.5 x 17.5 cm)
- Twelve ⅜-inch (9 mm) self-adhesive protective vinyl pads (These tiny dots for your clock's hour markers are typically used as appliance surface guards.)
- Approximately 12 inches (30 cm) of solid copper wire
- Clock movement and hands

TOOLS

- Work gloves
- Tinner's snips
- Artist's brush
- Double-sided tape
- Bent-nose pliers

INSTRUCTIONS

1 Wrap the aluminum flashing tightly around the shelving standard to create the base of the clock.

2 Using a point of the tinner's snips, poke holes through the flashing and into the slots in the standard where the shelf brackets (which form each side) will sit. Insert a bracket at each end of the standard, so the open bottoms of each bracket are facing each other (see figure 1).

3 Wedge the metal gutter guard in place between the brackets, and bend forward the excess gutter guard at the top.

4 There are two small holes at the upper end of each bracket. Cut a 4-inch (10 cm) piece of the 18-gauge copper wire, and thread it through the hole on one bracket, then through the gutter guard, and back through the hole. Twist the ends of the wire (which are now on the outside edge of the bracket), to secure them in place, creating decorative spiral loops with the wire. Repeat the process on the other side.

Figure 1

When You Need to Know Not Only When, But Where

One of the most useful applications of atomic clocks is the Global Positioning System (GPS), which lets people carrying a telephone-sized GPS receiver pinpoint their location anywhere on earth to within a few yards. The GPS system, completed in 1993 for military use, depends on 24 orbiting satellites carrying atomic clocks. The satellites, spread out so that a receiver can always "hear" four of them at once, continuously broadcast their position and the exact time, allowing the receiver to compute not only where it is but how long it will take to get somewhere else.

The usefulness of GPS has quickly spread into civilian territory: ship navigation, instrument landing at airports, crop dusting, mapping, surveying. Drivers of automobiles can find their way around strange cities. Backpackers can navigate mountainsides. Soon, using computerized "yellow pages," we'll be able to use GPS receivers to find a local restaurant, gas station, hospital, or hotel in any city or town in the world.

5 Paint the two wooden beads with metallic ruby paint, and let them dry.

6 Use the tinner's snips to cut a strip of rolled aluminum flashing to cover the front of the clock's base (the shelving standard). Paint the strip with random strokes and swirls of turquoise, copper, black, and metallic ruby paint.

7 Spray all the painted surfaces with acrylic sealer, and let them dry.

8 The rectangular piece of step flashing will become the clock face. On the two shorter sides of this piece, use the tinner's snips to poke a hole about 1 inch (2.5 cm) down from the top and ½ inch (1.3 cm) in from the side. In addition, poke a hole in the center that is large enough to accommodate the shaft of the clock movement.

9 Insert the clock movement, and attach it to the back of the clock face with two pieces of double-sided tape. Stick the 12 vinyl pads in place at the hour marks, and attach the hands to the movement.

10 On each bracket, there is a second small hole near the middle. You'll position the clock face between the brackets, with the holes you poked in the flashing lining up with the holes in the brackets. To do so, cut two 6-inch (15 cm) pieces of the 18-gauge copper wire. Thread one piece through the hole on the left side of the clock face and on through the hole in the left bracket, working the edge of the flashing into the opening on the inside edge of the bracket as you do. Twist the wire once to secure it, then thread a painted wooden bead onto the remaining wire on the outside of the bracket. Use the pliers to form decorative loops with the "dangling" wire. Repeat the process on the right side.

11 Create a spiral with the piece of solid copper wire and attach it to the gutter guard, just beneath the clock face, with 18-gauge wire.

12 Use double-sided tape to attach the painted strip of flashing to the front of the base of the clock.

Note: If your hand won't easily slip between the clock face and the gutter guard when you need to change the battery in the movement, simply loosen the wire on one side of the face to make the movement easier to reach.

Kid's Gift Clock

DESIGNER: LAUREN KRUCKE

No self-respecting grandparent, aunt, or uncle would refuse to hang your little darling's latest hand-drawn masterpiece on the refrigerator door. But just think how happy they'd be to receive something that offers function as well as budding artistic flair.

Any all-purpose bargain store will sell you an inexpensive kitchen clock. You know the look—colored plastic frame, plain white face with black numbers, clear plastic cover. Here are the simple steps for helping a young artist transform one into a personalized gift.

1 Pop off the plastic cover, unscrew the hands, and remove the paper clock face.

2 Typically, the back side of the clock face is blank. If so, you can simply flip it over. If not, use the face as a guide to cut a circle the same size out of sturdy white paper. Cut a hole in the middle of your cut-out circle.

3 Equip your resident artist with the supplies he or she needs to create a new clock face (crayons, markers, glitter, fabric scraps, beads, buttons, glue, etc.). You might want to provide some help when it comes to properly positioning the numbers.

4 Stick the new clock face in place over the movement, screw the hands back on, and snap the plastic cover back in place.

This simple project also makes a dandy activity for groups at events ranging from birthday parties to scout-troop gatherings.

A couple of original clocks designed by Lauren Krucke, age six, including an early work (bottom), designed when she was four, along with a recent piece. Lauren's grandparents are the proud owners of a retrospective of her hand-decorated clocks, created and presented as gifts over the years.

Plate Mosaic

YOU HAVE TO ADMIT, there's something irresistible about the fact that the first step in this project involves whacking breakable objects to pieces. All in the name of art, of course—the wreckage is raw material for a one-of-a-kind kitchen or dining room clock.

DESIGNERS: ALLISON AND TRACY PAGE STILWELL

MATERIALS

- Dishes for breaking
- Heavy plastic bag
- Plastic plate or platter, for base of mosaic
- Small trinkets or fun miniature kitchen items to work into mosaic
- 4 purchased figurines for number markers (To come up with their figurines, the designers purchased a set of small dip knives featuring miniature ceramic waiters as handles and removed the stainless steel spreaders.)
- Sheet of thick vellum or of the plastic material sold at craft stores for making stencils
- Black permanent marker
- Black acrylic paint
- Clock movement and hands

TOOLS

- Safety glasses
- Hammer
- Ruler
- Pencil
- Drill with ¼-inch (6 mm) drill bit
- Tile cutter (You can find a simple tile cutter in a craft store; it looks like a pair of pliers.)
- Industrial glue (available at craft stores)
- Grout
- Wet sponge
- Rubber gloves
- Scissors
- Artist's brush

INSTRUCTIONS

1 Put on the safety glasses, place your breakable dishes in the heavy plastic bag, and carefully break the dishes into small pieces with the hammer.

2 With the ruler, find the center of the plastic plate or platter and mark it with the pencil. Drill a ¼-inch (6 mm) hole in the center of the plate.

3 Experiment with the arrangement and design of your dish fragments on the plastic plate before you start gluing down your mosaic. Be playful—add jewels, marbles, or fun little kitchen pieces to enhance your design. Use the tile cutter to change the shape of the ceramic pieces or to nip off any sharp edges.

4 Glue the mosaic design and number-marker figurines in place and let them dry for at least 24 hours. Don't forget to leave the hole in the center open for the clock movement.

5 Once the glued mosaic is completely dry, follow the manufacturer's directions for applying the grout. Wearing rubber gloves, work the grout into all the spaces around the dish pieces. Let the grout dry for about 10 minutes, then wipe it off with a wet sponge. Repeat this process until you're happy with the results. Let the grouted mosaic dry overnight.

6 With the marker, draw shapes that resemble the ends of a spoon and a fork on the vellum or plastic stencil material (use actual utensils as guides, if necessary). You'll glue these plastic pieces onto your clock hands in the next step; make them small enough so that the clock hands will still be able to work properly and fit on your plate. Cut the spoon and fork shapes out with the scissors, then use the artist's brush to apply one or two coats of the black acrylic paint, covering the shapes completely.

7 Once the spoon and fork shapes are dry, glue them to the ends of the clock hands. Allow the hands to dry completely.

8 Insert the clock movement and attach the hands.

Tip: Craft supply stores often have fun assortments of small kitchen items, tiny tiles, jewels, and other eclectic pieces that can add personality and pizzazz to your mosaic clock.

Charmed, I'm Sure

TINY PLASTIC PTERODACTYLS, beads in the shape of bowling pins, little rubber fish, and shiny gems salvaged from long-forgotten pieces of costume jewelry. *Yes*, you've been telling people for years. *There is a reason I'm saving all this stuff*. Now it's clear to everyone just what you had in mind.

DESIGNERS: ALLISON AND TRACY PAGE STILWELL

MATERIALS

- Purchased clock base, tall arch style
- Acrylic paint in a bright color of your choice
- Large and varied assortment of charms, buttons, beads, and jewels (You can buy these supplies by the bag at craft stores.)
- Bezel clock movement to fit hole in base

TOOLS

- Artist's brush
- Medium-grade sandpaper
- Strong craft glue that dries clear

INSTRUCTIONS

1 Paint the surface of your clock base any color you like (the brighter the better is probably a fine approach). Let the paint dry completely, then sand the base as needed.

2 If you ever fell for the philosophy that less is more, let it go. Then, begin decorating your clock with charms and trinkets by gluing them onto the clock base. Piling pieces on top of one another and hanging some objects upside down only adds to the whimsy. Let the glue dry overnight. You may have to glue on your charms in stages, allowing one layer to dry before adding another.

3 Once you're satisfied with your design (such as when you're sure the clock won't hold one more piece of embellishment) and the glue is completely dry, insert the bezel movement in the base.

Vintage Pocketbook Clock

GATHER THE RIGHT INGREDIENTS, and you can put together a clock that looks like a set piece from a period play—and wins plenty of double takes. Simply copy this collection of high-society props, or start with the basic concept, then improvise away.

DESIGNER: TAMARA MILLER

MATERIALS

- Pocketbook with hard surface
- Assorted accessories (pair of ladies gloves, string of pearls, vintage sunglasses, etc.)
- Clock movement and hands

TOOLS

- Pencil
- Drill and bit to fit your clock movement shaft
- Hot glue gun and glue sticks

INSTRUCTIONS

1. Experiment to get a general idea of how you want to arrange your accessories on the pocketbook.

2. In the space remaining on the front of the pocketbook, mark where you want the center of the clock face to be, and drill a hole.

3. Insert the clock movement and attach the hands.

4. Arrange the accessories and glue them in place with the hot glue gun. In addition to gluing objects to the front of the pocketbook, you can also glue them to the inside and allow them to drape over the front edge.

I'd Rather Be Fishing

DESIGNER: TAMARA MILLER

ANY QUESTION WHAT'S *really* on the mind of the clock watcher who hangs this tell-tale timepiece near the computer screen? This is a clock that's also perfectly at home suspended from a rusty nail, amid all the hooks, lines, and sinkers it celebrates.

MATERIALS

- Rustic board (This designer used a board that measures approximately 7 x 18 inches [17.5 x 45 cm]. Select a board of any size, and simply adapt the instructions and materials to fit its dimensions.)
- Rope, enough to use as a hanger for your board
- Fishing net
- Floral wire
- Fishing-motif holiday ornaments
- Aquarium accessories
- Full-size fishing tackle and lures
- 12 small fishing lures or flies
- Clock movement and hands

TOOLS

- Drill and drill bits of various sizes
- Hot glue gun and glue sticks
- Wire cutters
- Ruler
- Pencil

INSTRUCTIONS

1. Drill holes in the top right and left corners of the board to accommodate the rope hanger. Insert the rope from the back of the board and knot each end. Fray the rope ends, if you want them to appear more worn.

2. On the right side of the board, arrange the fishing net until you like the look, and hot glue it in place.

3. Use the floral wire and glue gun to attach the fishing ornaments and aquarium accessories to the fishing net. Attach the full-size lures and flat decorative items to the board around the net with the glue gun, as well.

4. With the ruler and pencil, find and mark the center of the remaining space on the left side of the board. Drill a hole at the center mark to accommodate the clock shaft.

5. Position the 12 small lures and flies evenly around the clock face to represent the numbers, and hot glue them in place.

6. Insert the clock movement and add the hands.

Salvaged Cigar Box Clock

HERE'S THE STORY. A man stumbles onto a cigar box (one that's been aging under a stack of stuff in a garage for a couple of decades). Then (for no reason in particular) he dumps out the contents of his tool belt nearby. By the light of the full moon, the pieces spring to life and configure themselves into a clock suitable for a workshop, den, or hunting lodge. The how-to steps on the next page aren't quite as fantastic, but they're almost as easy.

DESIGNERS: ALLISON AND TRACY PAGE STILWELL

MATERIALS

- Wooden cigar box
- Variety of rusty nails, screws, etc.
- Found metal objects (The project shown features an old money clip, a mailbox latch, and a couple of decorative pieces.)
- Clock movement and hands

TOOLS

- Drill, 3/8-inch (9 mm) bit for the center hole, and a small bit (to fit your nails and screws)
- Clear-drying glue
- Instant iron and rust (Available at craft stores, these products work together to create a rusty effect.)

INSTRUCTIONS

1 Drill holes around the edge of the box. Place them randomly and at different angles.

2 Dip the sharp ends of your nails and screws into a bit of glue, then insert them in the holes.

3 Glue found objects at the 12, 3, 6, and 9 o'clock positions.

4 Paint the clock hands with the instant iron, then the instant rust, following the manufacturer's directions.

5 Insert the clock movement, and attach the hands once they're dry.

The Sound Heard (Somewhat Differently) Around the World

Mechanical clocks are operated by a series of different-sized moving wheels called a wheel train. To keep all the wheels turning in a controlled way once power is delivered, a mechanism called an escapement acts as a brake. Pins on the ends of a lever catch on the teeth of the escapement wheel, making the wheel train stop and start at regular intervals. When a pin catches, there's a tick. When it's released, there's a tock. Hence the tick-tock sound associated with clocks.

Well, tick tock to English speakers, that is. Turns out, the onomatopoeic phrase translates a bit differently, depending on the language you speak.

Arabic: *taktakah*
Catalan: *tic-tac*
Dutch: *tik tak* (which happens to be the name of a children's program on Belgian television, as well); also boem, boem, boem (pronounced "boom, boom, boom" but with a shorter "oo") for the deep, loud chimes of a grandfather clock, koekoek (pronounced "coocook") for a cuckoo clock, bim bam bom for church bells, and tring for an alarm clock
French: *tic tac* or *tac tac tac*
German: *das Ticktack*
Italian: *tic tac*
Japanese: *kachi kachi*, or *chakku tikku*

かちかち

Korean: *ttok-ttak*
Turkish: *tik tak* (pronounced "tick tuck")

Summertime (Living is Easy) Clock

DESIGNERS: ALLISON AND TRACY PAGE STILWELL

THOUGH TIME MAY FEEL AS IF IT'S FINALLY SLOWING down when you're swaying in a porch swing or puttering in the potting shed, it never actually stops. Here's a gentle reminder that it's time to stroll inside and slice the watermelon.

MATERIALS

- Straw hat, approximately 16 inches (40 cm) in diameter, with a wide, flat top
- 3 yards (2.7 m) of gauzy ribbon
- Dry or silk flowers, enough to decorate the brim of the hat
- Acrylic paint, olive green (to coat clock hands)
- Miniature garden tools, to embellish clock hands (These designers found miniature garden tools at a local gardening store. Nurseries, hardware stores, and card and gift shops may carry similar tiny tools.)
- Acrylic sealer
- Clock movement and hands

TOOLS

- Hot glue gun and glue sticks
- Artist's brush
- Craft glue
- Sharp scissors

INSTRUCTIONS

1 Using the hot glue gun, glue the length of ribbon around the brim of the hat. Leave the ribbon somewhat loose—this will allow you to manipulate it a bit as you add your flowers. Tie a big bow with the ends of the ribbon, and allow the remaining ribbon to dangle freely.

2 Arrange the larger flowers around the brim first. When you've settled on the arrangement, glue them in place. Add the smaller, accent flowers next, and glue them in place.

3 Apply one or two coats of olive green paint to the clock hands, and let them dry.

4 Remove the handles from the miniature tools, and use the craft glue to attach the heads of the miniature tools to the clock hands. Let the glue dry completely. Apply two coats of the acrylic sealer to the new clock hands to protect their surfaces.

5 Cut a small "X" in the center of the hat for your clock movement.

6 Insert the clock movement and attach the hands.

Working Lunch

CREATING ONE OF THESE FUNKY WORKING LUNCH BOXES is almost effortless. (If you can drill a hole, you can do this.) Want maximum impact? Line up a collection of 'em. They'd be perfect across the top of the kitchen cabinets or mounted over the back door—the one everyone uses each morning on their way to work and school.

DESIGNER: TAMARA MILLER

MATERIALS

- Metal lunch box
- 4 magnets, sized to fit around the clock hands
- Clock movement and hands

TOOLS

- Ruler
- Marker
- Drill with drill bit to fit your movement shaft

INSTRUCTIONS

1 Choose a metal lunch box without busy graphics to serve as the body of your clock.

2 Find and mark the center of the front of the lunch box, and drill a hole.

3 Insert the clock movement and attach the hands.

4 Position the magnets around the clock face to represent the numbers 3, 6, 9 and 12. Be sure to choose magnets that will fit around the clock face without obstructing the clock hands.

Tips: This is a perfect project to let kids help with once you've drilled the hole. It also makes for a versatile clock; you can easily switch the magnets every time a new holiday or season rolls around.

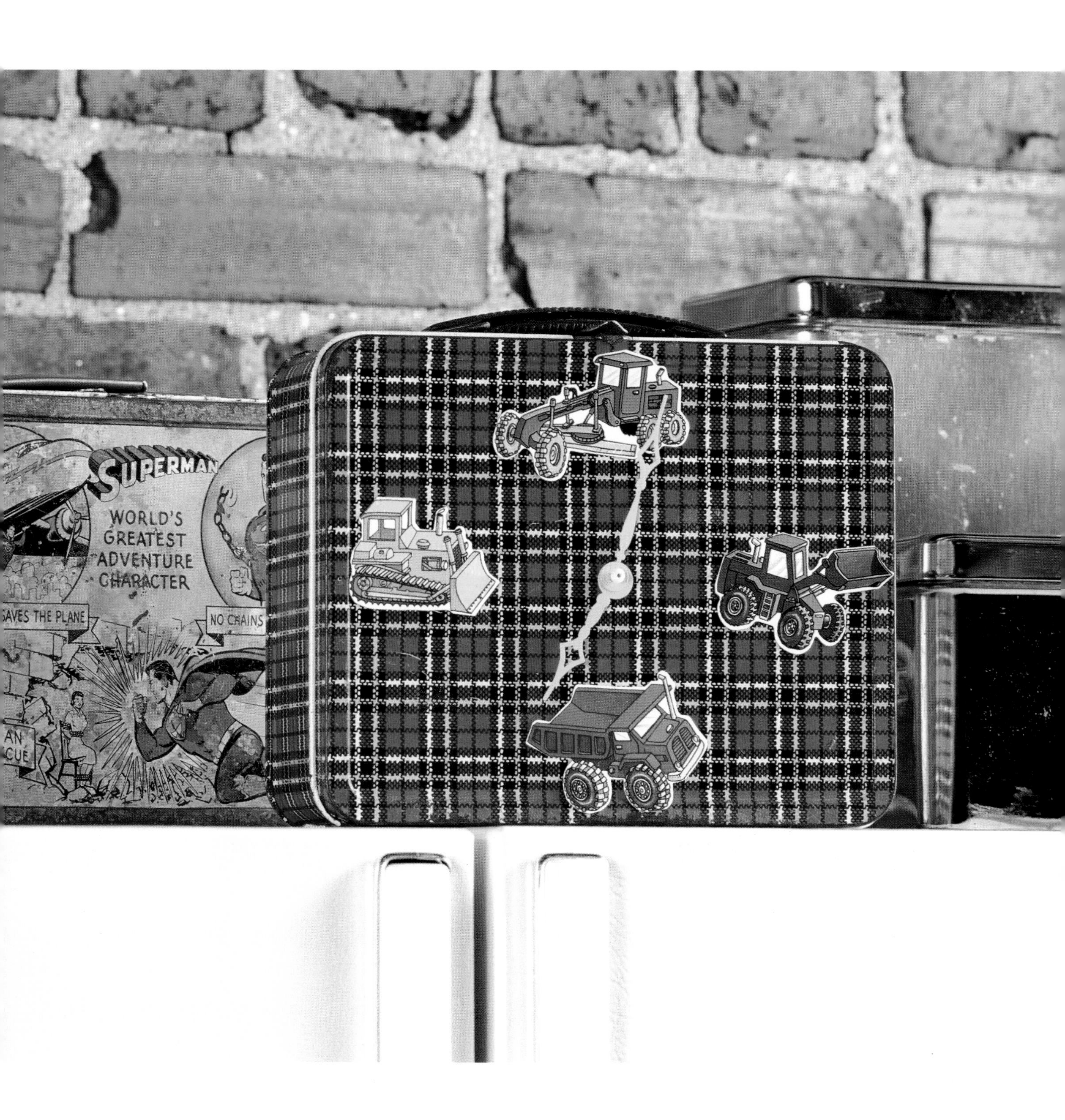
SUPERMAN
WORLD'S GREATEST ADVENTURE CHARACTER
SAVES THE PLANE
NO CHAINS

Clock of Habits

DESIGNER: TAMARA MILLER

THIS GAME BOARD turned clock is a tribute to habit-forming fun. Your call whether it's decorated with symbols of the habits on your to-kick list (check out the caffeine and fast cars shown here) or those you know you should develop (maybe tiny tokens in the shapes of fresh fruit and hiking boots?!).

MATERIALS

- Picture frame (This designer used a frame that measures approximately 16 x 19 inches [40 x 47.5 cm], with inner dimensions of 14½ x 11½ inches [36.3 x 28.8 cm]. But you can select a frame of any size for your clock—simply cut the cardboard and game board to fit.)
- Black spray paint
- 1 piece of thin cardboard, large enough to fit in frame
- 1 game board, large enough to fit in frame
- 12 assorted miniature novelty items
- Clock movement and hands

TOOLS

- Ruler
- Pencil
- Scissors or sharp craft knife
- Saw (optional)
- Drill and drill bit to fit your clock movement shaft
- Hot glue gun and glue sticks

INSTRUCTIONS

1 Spray paint the picture frame black, and let it dry completely. Apply more than one coat of spray paint, if necessary, for complete coverage.

2 Measure and cut a piece of cardboard to fit the dimensions of your picture frame. Insert the cardboard into the frame to serve as backing.

3 Make sure you have selected a game board that is sturdy enough to accommodate the items you plan to attach to the clock face. Measure and cut the game board to fit your frame, if necessary.

4 Use the ruler and pencil to find and mark the center of the game board. Drill a hole at the center point for the clock movement.

5 Arrange the 12 assorted miniature items around the clock face to mark the number locations. You can get both clever and creative with your number symbols. This designer selected a racing car decorated with a "1" to mark one o'clock, $7 worth of play money for the seven o'clock spot, etc.!

6 Once you're satisfied with your arrangement, hot glue the items in place.

7 Insert the clock movement and add the hands.

Terra-Cotta Saucer Clock

Designer: Terry Taylor

WARM WEATHER HITS, and you end up spending most of your time outside on the terrace or deck. But how do you know exactly how *much* time without a clock out there? Here's one designed to blend beautifully with all the potted flowers and plants.

MATERIALS

- 1 terra-cotta saucer, 14-inch (35 cm) diameter rim with 12-inch (30 cm) diameter flat bottom
- 6 or 8 chipped dinner plates in at least two colors of your choice (one color for the numerals and another for the background)
- Ceramic tile mastic
- Sanded floor grout in a color of your choice
- Hook or strong nail, for hanging
- Clock movement and hands

TOOLS

- Ruler
- Pencil
- Electric drill with ⅜-inch (9 mm) masonry drill bit
- Low-tack masking tape (wide painter's tape works well)
- Safety glasses
- Tile nippers
- Palette knife or disposable plastic knife (for spreading mastic on tile pieces)
- Small bucket or 6-cup (1.4 L) plastic container (for mixing grout)
- Plastic sheeting or newspaper to cover work area
- Latex gloves to protect hands
- Scraps of polyethylene foam wrap (white or green packaging material)

INSTRUCTIONS

1 Using the ruler and pencil, find the center of the saucer bottom and mark it.

2 Lightly mark the numeral positions of your clock face around the saucer bottom. This designer marked only numbers 3, 6, 9, and 12. If you'd like to mark them all, that's fine.

3 Use the electric drill equipped with the masonry bit to drill through the saucer at the center point. (Tip: When using a masonry bit, drill slowly, and don't put too much pressure on the bit. You may find it helpful to support the underside of the saucer with a scrap of wood to avoid putting too much pressure on the terra-cotta.)

4 Locate a place on the rim of the saucer above the number 12 spot and drill a hole for your hanging clock. (Don't count on hanging your clock with the hanger on the clock movement; the movement will be too far inside the saucer to reach the wall and will probably not be strong enough to support the weight of the terra-cotta saucer.)

5 With the masking tape, mask off the flat center section of the saucer (which will become your clock face) and the rim of the saucer, which won't be covered with mosaic.

6 Wearing safety glasses and using the tile nippers, nip off the rims of the plates. (This designer used only pieces of the plate rims, not the plate bottoms, to create the mosaic.) Simply nip the plate rims into large chunks. You can make them smaller to fit specific spots later.

7 Select the plate color you want to use for your numbers. (This designer chose to nip a particular design element—ivy leaves—to mark the number spots.) Use the tile nippers to make the pieces a uniform size and shape, if you like. (The number pieces shown here measure approximately 1 x 1¼ to 1½ inches [2.5 x 3.1 to 3.8 cm].)

8 Place the number pieces where they belong in the

The Beast: Mechanical Clock and Modern Myth

Maple Ridge, British Columbia, has a landmark clock that comes with its own tall tale—one its creator, Don Brayford, hopes will influence generations to come. The Beast Clock is a 12-foot-tall (3.7 m) hydraulically powered horse that rears its front legs and raises its tail to the sound of the Westminster Chimes each hour. When not in motion, the mechanical metal horse, unveiled in 1989, stands proudly atop a 24-foot (7.4 m) clock base, decorated with four clock faces, which are illuminated at night and decorated with lights every Christmas season.

But there's much more to The Beast than time and motion. In addition to dreaming up the clock (a secret project that kept Brayford and several workmen busy on nights and weekends for years), Don Brayford created an environmental fable to go along with it. His equestrian timekeeper plays a central role.

The legend of The Beast tells of a time when greedy people polluted the land and ignored its bounty. As Mother Nature spent the last of her energy repairing the damage, The Beast came thundering out of the valley in Maple Ridge, determined to punish those who had gained from the destruction of the environment. To this day, according to Brayford's story, the souls of those who polluted the earth for their own profit are imprisoned inside the body of The Beast, sentenced to look out at the world through its eyes. In support of the legend, The Beast's eyes are designed to move and roll from side to side when set in motion.

As his city changes and grows, Mr. Brayford says he hopes The Beast Clock will continue to serve as a community landmark—and that its mythic tale will be a reminder to all of the importance of caring for the environment.

clock face, positioning them flush with the outer edge of the saucer's flat bottom. Place the finished, rim edges of the pieces toward the center of the clock face. Use the tile mastic (applied with the palette knife or plastic knife) to adhere them to the saucer.

9 Continue to mosaic the clock face, using plate pieces in your second color. Arrange them in an even circle around the outer edge of the clock face, between the number pieces. Position them so that their finished rim edges face the center of the clock face. Leave approximately ⅛-inch (3 mm) of space between each piece of mosaic. If you choose to leave more, don't exceed ½ inch (1.3 cm). Use tile mastic to adhere them to the saucer.

10 Once you've finished the mosaic around the flat surface of the clock face, arrange and adhere plate pieces to the saucer rim. Cut the pieces to size with the tile nippers, and place them so that their rim edges face the top of the saucer rim (their broken edges should face the broken edges of the pieces on the flat surface). When you get to the hole you drilled in the rim, simply cut tiny bits of plate rim to place around the hole. After you've finished adhering all the plate pieces with tile mastic, allow the saucer to sit overnight.

11 Follow the manufacturer's instructions for mixing your tile grout. You probably won't need much more than ¾ cup (.2 L) of water to mix the grout for your clock.

12 Applying grout can be dirty work, so cover your work area with newspaper or plastic sheeting and wear some old clothes. Apply the grout to the tiled surfaces of the clock, using the foam wrap as a spreader. Remove excess grout as you go along. Allow the grout to set according to the manufacturer's instructions. Follow the manufacturer's clean-up instructions to remove grout haze.

13 Once your clock is thoroughly dry, remove the masking tape, insert the clock movement, and attach the hands.

14 To hang the clock, insert the hook or nail through the drilled hole in the rim at the top of the clock.

Gallery

A
Mary Engel: Watch Dog, 1998; 17" x 12" x 9" (43 x 30.5 x 23 cm); ceramic and mixed media. Photo: Walker Montgomery

B
Lynn Ludemann: Inside/Outside, 1994; 18" x 8" x 6" (46 x 20 x 15 cm); mixed metals. Photo: Lynn Ludemann

C
D. Lowell Zercher: Does Anybody Really Care?, 1996; 22.3" x 22" x 3" (57 x 57 x 7.5 cm); curly maple, curly Honduras mahogany, purple heart, glass, metal, India ink, paint, lacquer. Photo: Jack Holmquist

A

B

C

A

A
Robin C. Mangum, Jr: Squiggle-Mantle Clock, 1999; 27" x 20" x 5" (68.5 x 51 x 13 cm); stoneware, walnut, cherry. Photo: Paul Jerimias

B
Garry Knox Bennett: untitled, 1992; 26¼" x 19" x 10" (67 x 48.5 x 25.5 cm); found object, wood, metal, paint. Photo: M. Lee Fatherree

C
Lynn Ludemann: Aunt Flora, 1994; 37" x 18" x 16" (94 x 46 x 40.5 cm); mixed metals. Photo: Lynn Ludemann

C

B

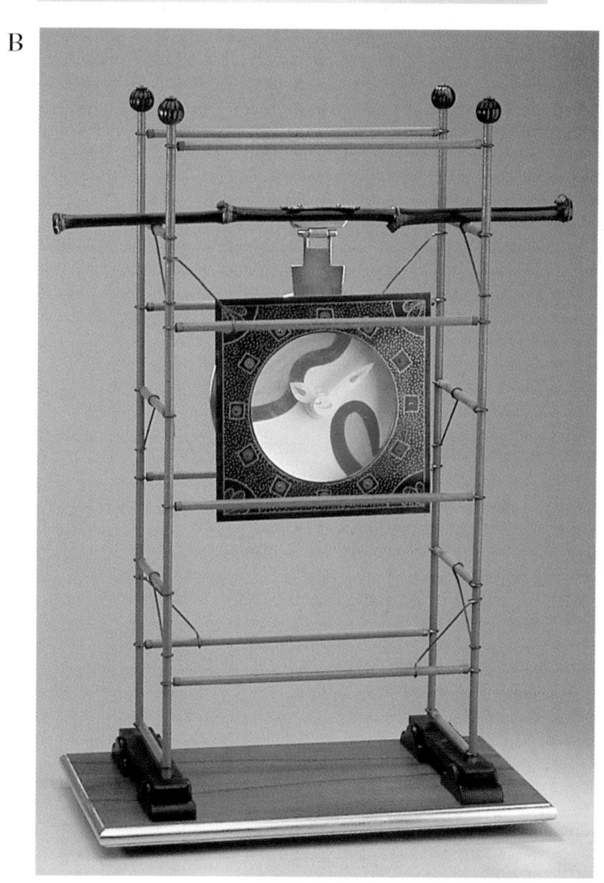

D

D
Garry Knox Bennett: Adam and Eve, 1976; 12" x 12" x 4" (30.5 x 30.5 x 10 cm); wood, brass, copper, cast pewter, bone. Photo: Nikolay Zurek

E
Thomas Roy Markusen: 420T Sculptural Clock, 1989; 48" x 10" x 5" (1 m x 25.5 cm x 13 cm); cast bronze, copper, and brass with black chrome and nickel electroplated finish. Photo: R. Margolis.

F
Brent Skidmore: Cairn Clock, 1999; 98" x 20" x 15" (2.5 m x 51 cm x 38 cm); basswood, pommele sapele. Photo: David Ramsey

F

E

A

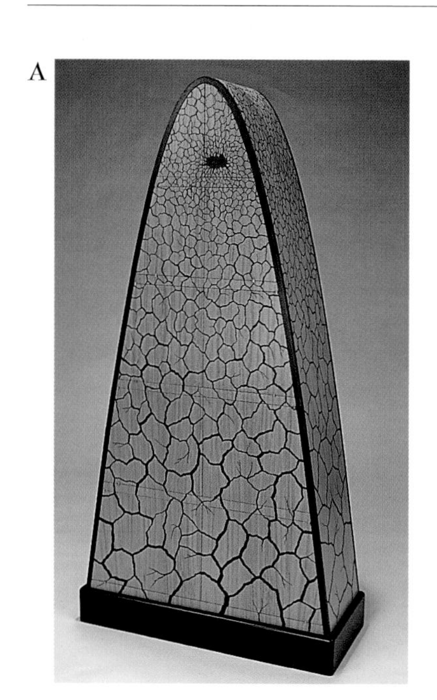

B

C

D

A
Silas Kopf: Parabola, 1992; 64" x 34" x 14" (1.5 m x 86.5 cm x 35.5 cm); satinwood and dyed black pearwood marquetry. Photo: David L. Ryan

B
Curtis K. LaFollette: Clock, 1999; 8½" x 4" x 5" (21.5 x 10 x 13 cm); sterling silver, brass, cast iron. Photo: Curtis K. LaFollette

C
D. Lowell Zercher: Big Guy, 1995; 42" x 19½" x 6" (1.1 m x 49.5 cm x 15 cm); Honduras mahogany, curly maple, cedar, bending poplar, brass, plastic, gold leaf, paints. Photo: Chris Arend

D
Penny Cash, Doug Hays: Time Table, 1999; 15" x 15" x 21" (38 x 38 x 53.5 cm); hand-forged steel and raku. Photo: Betsy Reed

E

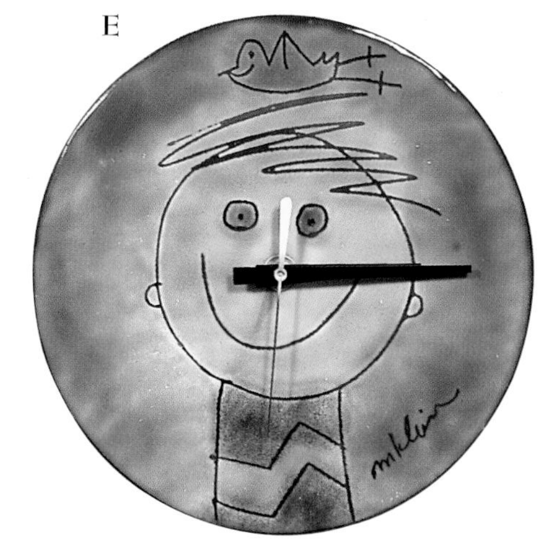

F

G

H

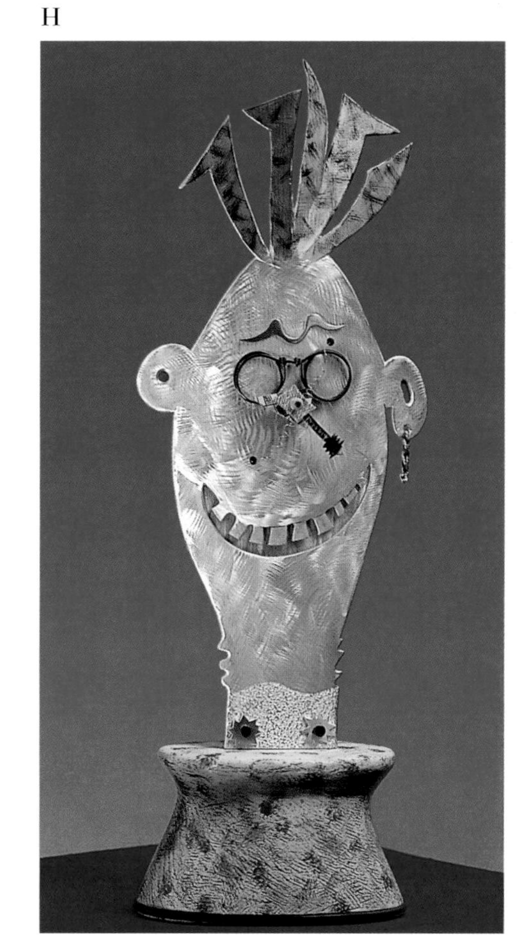

E
Mary Klein: Little Boy Clock, 1990; 9" (23 cm); enamel on copper. Photo: CPS Photography, St. Petersburg, FL

F
David Jerome Babb: I Wonder What Time He'll Be Home, 1995; 15" x 16" (38 x 40.5 cm); wood and acrylic.
Photo: D. Babb

G
Robert DuLong of Woodendipity, Inc.: Great Grandfather Clock, 1995; 6'8" x 19" x 15" (2 m x 48.5 cm x 38 cm); red cedar and pine.
Photo: Gerry Young

H
D. Lowell Zercher: Rock On, 1998; 17" x 7" x 3" (43 x 18 x 7.5 cm); aluminum, wood, brass, wire, plastic, and paint.
Photo: Chris Arend

PATTERNS AND TEMPLATES

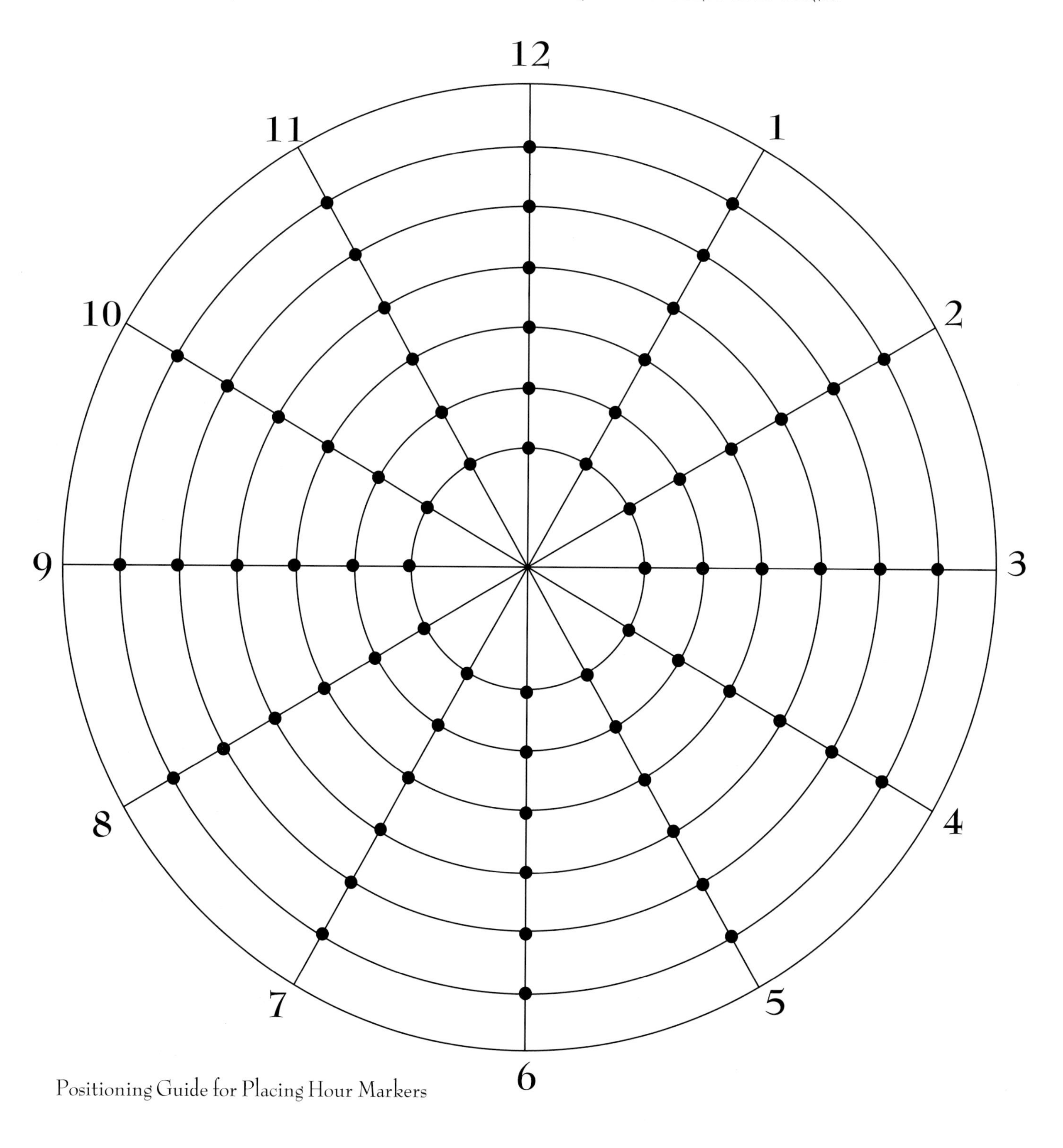

Positioning Guide for Placing Hour Markers

The Clock That Time Undid

Wired

Flying Furniture Clock
Paper Moon
Brown Bagging It
I'm Late, I'm Late!

CONSULTING DESIGNERS

Allison and **Tracy Page Stilwell** are creative and clever sisters who love to work with cloth and threads to form contemporary quilts and dolls, using a serendipitous artistic process that brings a new level of freedom of expression to these old crafts. Their art quilts are full of brilliant bursts of color and texture. The dolls are creatures from a world of their own.

Emerging from traditional artistic roots, Tracy now incorporates political and personal messages into her unpredictable combination of cloth, paint, wood, beads, bones, polymer clay, paper-based clay, rope, twigs, roots, yarns, threads, and found objects. Her work has been featured in a wide variety of books and magazines, and it has led to teaching and curating opportunities.

Allison, the taller, younger sister, feels that her greatest gifts have been her creative freedom and her ability to encourage others to embrace their own. As an imaginative crafts designer, her work has appeared in national magazines such as *Family Circle*, *Woman's Day*, and *Better Homes and Gardens*, and in many craft books. She has also authored a book of inventive designs for painting children's furniture.

Knowing how the inner critic can keep creativity at bay, as teachers, the Stilwells encourage others to take risks, to be brave, and to defy the pressure of perfectionism...pushing students to trust their own vision and use of materials.

In a recent turn, both have taken on the study of computer graphics, diving into the techno-cyber world of bits, bytes, and HTML code. Though they sometimes feel as if this new frontier is changing faster than their synapses are snapping, they are forging ahead with confidence (most of the time), and hope to combine their eye for color and design with the possibilities of the keyboard.

Practicing truth and gratitude, staying with the stitching, keeping the machine oiled, daily practice and patience with the computer, and snacking for strength, they continue down the path of their somewhat remarkable and very full lives.

CONTRIBUTING DESIGNERS

Bruce Allison and **Jill Mayberg** create clocks and other decorative items for City of Clay, a division of Paper-Wing Studios in Oakland, California.

Barbara Bussolari is a retired Massachusetts schoolteacher now living in the mountains of western North Carolina. For the past 20 years, she has designed, made, and sold cards using handmade and hand-decorated papers and handwoven material. She is now exploring and expanding into paper jewelry and dolls made with handwoven materials and handmade papers.

Irene Semanchuk Dean gets inspiration for her polymer clay work from many sources, most recently nontraditional quilts. She is a member of the National Polymer Clay Guild and one of the instigators of the Blue Ridge Polymer Clay Guild. Irene cites the generous and sharing nature of polymer clay enthusiasts as one of the nicest things about the medium. She designs and creates her work in her studio in Weaverville, North Carolina, and presents it under the name Good Night Irene! at regional art and craft shows and in specialty shops. She is currently writing a book on polymer clay for Lark Books. You can see her work at www.pobox.com/~ good.night.irene and contact her at good.night.irene@pobox.com.

Kimberly Hodges maintains a full-time production and design studio in Asheville, North Carolina. Her company, Queen's Crescent, creates a colorful line of home accents for more than 500 stores and galleries nationwide.

Rolf Holmquist is a Swedish-born artist known for his specialty, medical embossings. He has earned more than 100 awards for his artistic interpretations, which are punctuated with found objects such as actual medical devices, x-rays, and graphs.

Lynn B. Krucke lives in Summerville, South Carolina, with her husband and daughter. She has long been fascinated with handcrafts of all types, and her favorite projects incorporate elements from more than one craft.

Claudia Lee is a full-time paper-maker and currently the resident artist in fibers at the Appalachian Center for Crafts in Smithville, Tennessee.

Shelley Lowell is an award-winning graphic designer, illustrator, and fine artist. Her paintings and sculpture have been exhibited in museums and galleries in many cities throughout the United States. She resides in Alexandria, Virginia, with her three cats, none of whom are the ones depicted on the clock!

Tamara Miller is a mom first and crafter second, who lives in Hendersonville, North Carolina, with her husband, Jeff, and son, Beck. She is especially motivated by projects that allow her to incorporate the interests of her family while utilizing her creativity.

Jean Tomaso Moore is a part-time multimedia artist who has been creating art in one form or another for as long as she can remember. She lives with a humble and patient husband in the beautiful hills of Asheville, North Carolina. Contact her at LeaningTowerArt@aol.com.

Now living "the good life" on four and a half acres of quiet land outside of Lincoln, Illinois, retired carpenter **Don Shull** spends summers in his flower garden and winters in his workshop. These seasonal crafts complement each other well, he says. Summers allow him to observe the garden bugs he describes as "fascinating," while colder weather affords him time to recreate them in everything from whirligigs and mailboxes to, recently, clocks!

Terry Taylor is a multitalented artist whose work ranges from beading and lamp making to gilding and the pique-assiette mosaic technique. He allows bits and pieces of many art forms to influence his one-of-a-kind creations.

Kim Tibbals-Thompson lives in Waynesville, North Carolina. She is a frequent contributor to craft books and enjoys drawing, sewing, gardening, herbal crafting, and broom making. By day, she is a graphic designer.

GALLERY ARTISTS

David Babb wanders around bumping into things and getting occasional flashes of creative brilliance when he's not busy fighting environmental terrorists and saving the free world.

Garry Knox Bennett, a California furniture maker, often designs one-of-a-kind clocks which he exhibits independently or along with his furniture pieces.

Robert "Geppetto" DuLong creates an array of functional folk art sculptures for the home and garden. His work is shown at the website www.woodendipity.com, and marketed by his Guild of Freelance Artists and Woodendipity, Inc.

Mary Engel received her MFA from the University of Georgia. Her work has been the subject of several museum exhibits. In Atlanta, Georgia, Engel's work is featured at the Marcia Wood Gallery.

Artists **Doug Hays** and **Penny Cash** design and market their hand-forged steel and raku clocks through their business, Hays-Cash Designs.

Mary Klein is a studio artist living and working in St. Petersburg Beach, Florida. She is known internationally for her enamel portraits, and her work is included in museum collections around the world.

Silas Kopf designs and builds furniture with a focus on marquetry decoration.

Curtis K. LaFollette is a silversmith who specializes in utilitarian hollowware.

Lynn Ludemann creates and markets sculptured clocks featuring vintage and recycled metals through her business, Lynn Ludemann.

Rob Mangum, a second-generation potter from North Carolina, works with his wife, Beth, at their studio, Mangum Pottery. Rob's work has evolved to include ceramic and mixed media furniture.

Brent Skidmore, of Charlotte, North Carolina, specializes in unique sculptures, clocks, tables, mirrors, lamps, chests, chairs, and more.

Master metalsmith **Thomas Markusen** designs and creates distinctive decorative accessories using copper with patinas at Markusen Metal Studios, Ltd., in Kendall, New York.

D. Lowell Zercher's clocks range from whimsical to elegant and are included in both private and public collections.

ACKNOWLEDGMENTS

Who has the time (not to mention all the necessary expertise and resources) to tackle the topic of clocks alone? Not us. Thanks to the many who contributed to this book.

- ➢ Craft-supply manufacturer Walnut Hollow, for donating clock supplies used by several designers and in many of the shots in the book's Basics section.
- ➢ Chris Bryant and Skip Wade, for opening their beautiful home to our cameras.
- ➢ The following individuals, for taking the time to translate "tick tock" into a host of foreign languages: Inge Van der Cruysse-Van Antwerpen, Melinda S. Bayne, Erika H. Gilson, Stephen Harroff, Young-Key Kim-Renaud, Josep Miquel Sobrer, Nancy Virtue, and Masato Yabe.
- ➢ The Natural Home in Asheville, North Carolina, for allowing us to pick and choose photo-shoot props from their fabulous store.
- ➢ Aloha Tower Marketplace, Honolulu, Hawaii, for providing photos of the historic Aloha Tower, page 25.
- ➢ Hönes GmbH, Germany, for providing the photos of cuckoo clocks on page 53.
- ➢ Mr. Don Brayford of Pitt Meadows, British Columbia, Canada, designer and maker of The Beast Clock, page 116, for information and photographs.
- ➢ Clock and Watch Museum Beyer Zurich, for providing historical images of clocks from their collection, pages 7–11.
- ➢ David McGan (who "enjoys exploring new lands, seeking the unusual and unique"), for contributing his personal photos of the house-size cuckoo clock, page 53.
- ➢ The Niagara Parks Commission in Niagara Falls, Ontario, Canada, for information on and photographs of the Niagara Floral Clock, page 31.
- ➢ Yatin Chawathe of Berkeley, California, for photographs of the Westminster Clock, page 36. Additional photos are featured on his website: www.cs.berkeley.edu/~ yatin/tmp/bigben/.

INDEX